D1338164

Newnes Data Communications Pocket Book

Newnes
Data Communications
Pocket Book

Fourth edition

Michael Tooley
Steve Winder

OXFORD   AMSTERDAM   BOSTON   LONDON   NEW YORK   PARIS
SAN DIEGO   SAN FRANCISCO   SINGAPORE   SYDNEY   TOKYO

Newnes
An imprint of Elsevier Science
Linacre House, Jordan Hill, Oxford OX2 8DP
225 Wildwood Avenue, Woburn, MA 01801-2041

First published 1989
Reprinted 1990
Second edition 1992
Reprinted 1993, 1994, 1995
Third edition 1997
Reprinted 1998 (twice), 1999
Fourth edition 2002

British Library Cataloguing in Publication Data
A catalogue record for this book is available from the British Library

ISBN 0 7506 52977

For information on all Newnes publications visit our website at
www.newnespress.com

Typeset by Laserwords Private Limited, Chennai, India
Printed and bound in Great Britain

Contents

Preface

Data communications continues to expand due to the increased use of multi-media computers and through the use of the Internet and company-wide Intranets. The amount of data traffic carried over public telecommunication networks now exceeds that of voice traffic. Data communications links range from a simple low-speed modem operating over a pair of copper wires, through to complex packet switched networks operating over an optical fibre.

'Data' could be defined as non-real-time digital information such as data, photographic and video files. However, it could now also include real-time video streams and voice traffic since these are digitised and can have similar characteristics to data traffic. The convergence of all telecommunications traffic into packet based transmission such as Internet Protocol (IP) blurs the distinction between real-time and data traffic even more. The main distinction between them is the time delay in transporting the information from the source to the recipient; voice and real-time video must not be unduly delayed.

This fourth edition of the *Data Communications Pocket Book* attempts to briefly describe all current forms of data communications, from computer interfaces and cables through to the protocols used in packet based networks. New material includes Universal Serial Bus (USB) and Firewire interfaces, as well as CAT-5 cables and Internet Protocol version 6 (IPv6). Some material from the third edition has been removed and the remaining topics have been updated. As with any small book, there is never enough space to publish all the information that may be needed. However, this book will hopefully contain enough information to help engineers and technicians whilst working away from their bulky reference books.

Steve Winder

1 Glossary

Abbreviations commonly used in data communications

AAL	asynchronous transfer mode adaptation layer
AAT	arbitrated access timer
ABM	asynchronous balanced mode
ABR	available bit rate
AC	access control
AC	alternating current
ACD	automatic call distributor
ACF	advanced communication function
ACIA	asynchronous communications interface adaptor
ACK	acknowledge
ACU	auto-call unit
ADCCP	advanced data communication control procedure
ADLC	add-on data link control
ADPCM	adaptive pulse code modulation
ADSL	asymmetrical digital subscriber line
AF	audio frequency
AFP	AppleTalk file protocol
ALOHA	(an experimental radio broadcast network)
AM	amplitude modulation
AMI	alternate mark inverted
ANI	automatic number identification
API	application program interface
APPC	advanced program-to-program communication
ARC	attached resources computing
ARM	asynchronous response mode
ARO	automatic request for repetition
ARP	address resolution protocol
ARPANET	Advanced Research Projects Agency Network
ARQ	automatic request for retransmission
ASCII	American standard code for information interchange
ASK	amplitude-shift keying
ASR	automatic send/receive
ATDM	asynchronous time division multiplexing
ATM	asynchronous transfer mode

BBS	bulletin board system
BCC	block check character
BCD	binary coded decimal
BCS	binary synchronous communications
BDLC	Burroughs data link control
BERT	bit error rate test
BIOS	basic input/output system
BISDN	broadband integrated services digital network
BLERT	block error rate test
bps	bits per second
BRI	basic rate interface
BSC	bisynchronous communications
BSE	basic service element
C7	see SS7
CANTAT	Canada transatlantic telephony cable
CASE	common applications service elements
CATV	community antenna television (ie, cable TV)
CBDS	connectionless broadband data service
CBR	constant bit rate
CBX	computerised branch exchange
CC	control codes
CCP	communications control program
CCS	common-channel signalling
CCU	communications control unit
CD	carrier detect
CDMA	code division multiple access
CDP	conditional di-phase
CEPT	European conference of Postal and Telecommunication Administrations
CFR	Cambridge fast ring
CHI	communications hardware interface
CICS	customer information control system
CILE	call information logging equipment
CMIP	common management information protocol
CMOS	complementary metal oxide semiconductor
CNM	communications network management
CO	central office
CODEC	coder-decoder
CPE	customer premises equipment
cps	characters per second
CPU	central processing unit
CRA	call routing apparatus

CRC	cyclic redundancy check
CRT	cathode ray tube
CSMA	carrier sense multiple access
CSMA/CA	CSMA with collision avoidance
CSMA/CD	CSMA with collision detection
CSPCN	circuit-switched public data network
CSU	channel service unit
CTA	circuit terminating equipment
CTS	clear to send
CUG	closed user group
CVSD	continuously variable slope delta modulation
DA	destination address
DAA	data access arrangement
DACS	digital access and cross-connect system
DART	dual asynchronous receiver/transmitter
DASS	digital access signalling system
dB	decibel
dBm	decibels relative to a reference level of 1 mW
DC	direct current
DCD	data and carrier detect
DCE	data circuit-terminating equipment
DCE	data communications equipment
DDCMP	digital data communication message protocol
DDD	direct distance dialling
DDI	direct dial-in
DDN	digital data network
DDS	Dataphone digital services
DDS	digital data service
DEA	data encryption algorithm
DECT	digital European cordless telephone
DES	data encryption standard
DID	direct inward dialling
DNIC	data network identification code
DOV	data over voice
DPNSS	digital private network signalling system
DPSK	differential phase-shift keying
DQDB	distributed queue dual bus
DRS	data rate select
DSA	distributed systems architecture
DSB	double sideband
DSBSC	double sideband suppressed carrier
DSC	district switching centre

DSL	digital subscriber line
DSLAM	digital subscriber line access multiplexer
DSU	digital service unit
DTE	data terminal equipment
DTMF	dual tone multi-frequency
DTR	data terminal ready
DUP	data user part
DXI	data exchange interface
EBCDIC	extended binary coded decimal interchange code
EBX	electronic branch exchange
ED	ending delimiter
EDI	electronic data interchange
EDU	error detecting unit
EFT	electronic funds transfer
EISA	extended industry standard architecture
ELR	earthed loop
EMA	enterprise management architecture
EMC	electromagnetic compatibility
EMI	electromagnetic interference
ENQ	enquiry
EOT	end of transmission
EPoS	electronic point of sale
EPSS	experimental packet switching service
ESF	extended superframe format
ETB	end of transmitted block
ETS	European Telecommunications Standard
ETX	end of text
FAX	facsimile
FC	frame control
FCS	frame check sequence
FDDI	fibre distributed data interface
FDM	frequency division multiplexing
FE	format effectors
FEC	forward error control
FEP	front end processor
FIFO	first-in, first-out (memory)
FM	frequency modulation
FS	frame status
FSK	frequency-shift keying
FTAM	file transfer access and management
FTP	file transfer protocol

FTTC	Fibre to the curb
FTTH	Fibre to the home
FXO	foreign exchange office
FXS	foreign exchange subscriber
GHz	10^9 Hz
GND	ground
GOSIP	Government OSI profile
GSC	group switching centre
GSM	global system for mobile
GUI	graphical user interface
HDB3	high-density bipolar code no. 3
HDLC	high-level data link control
HDSL	high bit rate digital subscriber line
HDTV	high-definition television
HF	high frequency
HM	hybrid modulation
HSLN	high-speed local network
HTML	hypertext mark-up language
Hz	Hertz (cycles per second)
IA5	international alphabet no. 5
ICMP	Internet control message protocol
ICP	interconnection protocol
IDA	integrated digital access
IDD	international direct dialling
IDN	integrated digital network
IEC	inter-exchange carrier
ILD	injector laser diode
ILEC	incumbert local exchange carrier
IMP	interface message processor
INFO	information
I/O	input/output
IOT	inter-office trunk
IP	Internet protocol
IPMS	interpersonal message processor
IPSS	international packet-switched service
IPX	Internet packet exchange
IRQ	interrupt request
IS	information separator
ISD	international subscriber dialling
ISDN	integrated services digital network

ISN	information systems network
ISP	Internet service provider
ISPBX	integrated services private automatic branch exchange
IT	information technology
ITA2	international telegraph alphabet no. 2
ITC	independent telephone company
ITU	International Telecommunications Union
IVDT	integrated voice and data terminal
JPEG	Joint Photographic Experts Group
JTMP	joint transfer and manipulation protocol
kHz	kilohertz
KTS	key telephone system
LAM	line adaptor module
LAN	local area network
LAP	link access protocol
LAPB	link access protocol balanced
LAPM	link access procedure for modems
LAT	local area transport
LATA	local access and transport area
LCD	liquid crystal display
LD	loop disconnect
LDM	limited distance modem
LEC	local exchange carrier
LED	light emitting diode
LEO	low earth orbit
LF	low frequency
LLC	logical link control
LMDS	local multipoint distribution service
LRC	longitudinal redundancy check
LSB	lower sideband
LSI	large scale integration
LT	line termination
LTE	line terminating equipment
LU	logical unit
LWT	listen while talk
MAC	medium access control
MAN	metropolitan area network
MAP	manufacturing automation protocol
MAU	multi-station access unit
MCA	micro-channel architecture

MCVF	multi-channel voice frequency
MF	medium frequency
MF	multiple frequency
MHS	message handling system
MHz	megahertz
MIB	management information base
MIPS	million instructions per second
MIS	management information system
MNP	Microcom network protocol
MODEM	modulator de-modulator
MPLS	multi-protocol label switching
MPEG	Moving Picture Experts Group
MPX	multiplexer
MSC	main switching centre
MSN	Microsoft Network
MTA	message transfer agent
MTBF	mean time between failure
MTTF	mean time to failure
MTTR	mean time to repair
MTU	maximum transmission unit
MUX	multiplexer
NAK	negative acknowledgement
NAU	network addressable unit
NCC	network control centre
NCOP	network code of practice
NCP	network control program
NCP	network core protocol
NDIS	network driver interface specification
NETBIOS	network basic input/output system
NFS	network file server
NFS	network file system
NIFTP	network-independent file transfer protocol
NITS	network-independent transport service
NLM	NetWare loadable module
NMP	network management protocol
NMU	network management unit
NNTP	network news transport protocol
NOC	network operations centre
NORE	nominal overall reference equivalent
NRM	normal response mode
NRZ	non-return to zero

NRZI	non-return to zero inverted
NT	network termination
NT1	network termination no. 1
NTE	network terminating equipment
NTSC	National Television Standards Committee
NTU	network terminating unit
NUA	network user address
OC	optical carrier
OC3	155 Mb/s data over fibre
OCR	optical character recognition
ODI	open data link interface
ODI	optical data link interface
ONU	optical network unit
OPT	open protocol technology
OSI	open systems interconnection
PABX	private automatic branch exchange
PAD	packet assembler/disassembler
PAM	pulse amplitude modulation
PAP	packet level procedure
PAT	priority access timer
PAX	private automatic exchange
PBX	private branch exchange
PC	personal computer
PCI	pre-connection inspection
PCM	pulse code modulation
PCN	personal communications network
PDA	personal digital assistant
PDN	public data network
P/F	poll/final
PM	phase modulation
PM	pulse modulation
PMBX	private manual branch exchange
PMR	private mobile radio
PON	passive optical network
POP	point of presence
POS	point of sale
POTS	plain old telephone service
PPM	pulse position modulation
PPS	pulses per second
PRI	primary rate interface

PSDN	packet switched data network
PSE	packet switching exchange
PSK	phase-shift keying
PSN	packet switching network
PSPDN	packet switched public data network
PSS	packet switched service
PSS	Packet SwitchStream
PSTN	public switched telephone network
PSU	power supply unit
PTO	public telecommunications operator
PTT	post, telegraph and telephone
PU	physical unit
PUC	public utilities commission
PVC	permanent virtual circuit
PWM	pulse width modulation
QAM	quadrature amplitude modulation
QPSK	quadrature phase-shift keying
QSAM	quadrature sideband amplitude modulation
RAM	random access memory
RBT	remote batch terminal
RC	receive clock
RD	receive data
REJ	reject
RF	radio frequency
RFI	radio frequency interference
RFS	ready for sending
RFS	remote file service
RI	ring indicator
RJE	remote job entry
RMON	remote monitoring device
RNR	receiver not ready
RO	receive only
ROM	read-only memory
RPC	remote procedure call
RR	receiver ready
RS	recommended standard
RT	resynchronisation timer
RTS	request to send
RU	request unit
RU	response unit
RZ	return to zero

SA	source address
SAA	systems application architecture
SAP	service access point
SCRA	single-line call-routing apparatus
SCSI	small computer system interface
SCTS	secondary clear to send
SCVF	single-channel voice frequency
SD	starting delimiter
SDCD	secondary data carrier detect
SDH	synchronous data heirarchy
SDLC	synchronous data link control
SDSL	symmetrical digital subscriber line
SFDM	statistical frequency division multiplexing
SFT	system fault tolerance
SG	signal ground
SIO	serial input/output
SMB	server message block
SMDS	switched multi-megabit data service
SMTA	single-line multi-extension telephone apparatus
SMTP	simple mail transfer protocol
S/N	signal-to-noise ratio
SNA	systems network architecture
SNADS	systems network architecture distribution services
SNDCF	subnetwork-dependent convergence facility
SNICF	subnetwork-independent convergence facility
SNMP	simple network management protocol
SNR	signal-to-noise ratio
SOH	start of heading
SONET	synchronous optical network
SPC	stored program control
SPX	sequenced packet exchange
SQ	signal quality
SQL	structured query language
SRD	secondary receive data
SRTS	secondary request to send
SS	signalling system
SS7	signalling system no. 7
SSB	single sideband
SSBSC	single sideband suppressed carrier
SSCP	system services control point
STA	spanning tree algorithm
STD	secondary transmitted data

STD	subscriber trunk dialling
STDM	statistical time division multiplexer
STM	statistical multiplexer device
STP	shielded twisted pair
STS	space-time-space
STS	synchronous transport signal
STX	start of text
SVC	switched virtual circuit
SYN	synchronous idle
TA	terminal adapter
TACS	total access communications system
TAN	trunk access node
TAPI	telephony application programming interface
TASI	time assignment speech interpolation
TBR	timed break
TC	transmit clock
TCAM	telecommunications access method
TCM	trellis code modulation
TCP	transmission control protocol
TCP/IP	transmission control protocol/Internet protocol
TCT	toll connecting trunk
TD	transmitted data
TDM	time division multiplexing
TDMA	time division multiple access
TDR	time domain reflectometry
TE	terminal equipment
TFTP	trivial file transfer protocol
TIC	token ring interface coupler
TIP	terminal interface processor
TJF	test jack frame
TOP	technical and office protocol
TRIP	transfer rate of information bits
TSE	terminal-switched exchange
TST	time-space-time
TTP	transaction tracking system
TXE	electronic exchange
TXK	crossbar exchange
UA	user access
UART	universal asynchronous receiver/transmitter
UDP	user datagram protocol
UHF	ultra high frequency

UNI	user-network interface
UNMA	unified network management architecture
USART	universal synchronous/asynchronous receiver/transmitter
USB	universal serial bus
USB	upper sideband
UTP	unshielded twisted pair
VADS	value-added data service
VAN	value added network
VANS	value-added network service
VC	virtual circuit
VCI	virtual channel identified
VDSL	very high bit rate digital subscriber line
VDT	video display terminal
VDU	visual display unit
VHF	very high frequency
VIP	VINES Internet protocol
VIPC	VINES interprocess communications protocol
VPI	virtual path identifier
VPN	virtual private network
VRC	vertical redundancy check
VSB	vestigial sideband
VTAM	virtual telecommunications access method
VTP	virtual terminal protocol
WAN	wide area network
WATS	wide area telecommunications service
WF	wait flag
XNS	Xerox network services
XTC	external transmit clock

Glossary of data communications terms

Acknowledgment
A signal which indicates that data has been received without error.

Address
A reference to the location of the source or destination of data. Each node within a network must be given a unique numeric identifying address.

Adaptive differential pulse code modulation
CCITT standard for encoding analog voice signals into digital form at 32 kbps (ie, half the standard PCM rate).

Alternate mark inversion
Bipolar coding system in which successive 1s (ie, *marks*) alternate in polarity.

Alternating mode
Half-duplex (ie, alternate send/receive) operation.

Amplifier
Circuit or device which increases the power of an electrical signal.

Amplitude
Peak excursion of a signal from its rest or mean value (usually specified in volts).

Amplitude modulation
A modulation method in which the amplitude of a carrier is modified in accordance with the transmitted information.

Analog loopback
A method of testing an item of data communications equipment in which outgoing analog signal (the *line signal*) is connected back to the analog input of the device and a comparison made (see also *digital loopback*).

Analog signal
A signal that can vary through an infinite number of amplitude levels (see also *digital signal* and *analog transmission*).

Analog transmission
Method of transmission in which information is conveyed by analog (eg, sinusoidal) signals.

Application layer
The top layer of the ISO model for OSI.

Asymmetrical digital subscriber line
A transmission system used to carry broadband signals over a copper pair.

Asynchronous transfer mode
Packet switching technique that uses fixed length packets of data (*cells*) sent at arbitrary intervals of time (note that, within the cell, the timing of bits is synchronous with a clock signal).

Asynchronous transmission
Transmission method in which the time between transmitted characters is arbitrary. Transmission is controlled by *start and stop* bits and no additional synchronising or timing information is required.

Attenuation
Decrease in the magnitude of a signal (in terms of power, voltage or current) in a circuit.

Balanced
In an *electrical* context a balanced line is one in which differential signals are employed (ie, neither of the conducting paths is returned to earth). In the context of the *data link layer* a balanced protocol is one involving a peer relationship of equal status (ie, not master–slave).

Balanced line
A balanced line is one in which the voltages on the two conductors are equal in magnitude but of opposite polarity. Neither of the two conductors is at ground potential. An example of a balanced line is a *twisted pair* (see also *unbalanced line*).

Band splitter
A multiplexer which divides the available bandwidth into several independent sub-channels of reduced bandwidth (and consequently reduced data rate when compared with the original channel).

Bandwidth
Range of frequencies occupied by a signal or available within a communication channel. Bandwidth is normally specified within certain defined limits and can be considered to be the difference between the upper (maximum) and lower (minimum) frequencies within the channel.

Baseband
The range of frequencies occupied by a digital signal (unchanged by modulation) which typically extends from d.c. to several tens or hundreds of kilohertz depending upon the data rate employed.

Baseband LAN
A local area network which employs baseband transmission techniques.

Baseband transmission
Transmission method in which digital signals are passed, without modulation, directly through the transmission medium.

Baud
A unit of signalling speed expressed in terms of the number of signal events per second.

Baud rate
Signalling rate (note that this is not necessarily the same as the number of bits transmitted per second).

Baudot code
A code used for data transmission in which each character is represented by five bits. Shift characters are used so that a full set of upper and lower case letters, figures and punctuation cannot be transmitted.

Binary synchronous communication
IBM Communication protocol which employs a defined set of control characters and control sequences for synchronised transmission of binary coded data (often referred to as *bisync*).

Bit
A contraction of 'binary digit'; a single digit in a binary number.

Bit error rate
A measure of the number of errors produced in a data communications systems. Bit error rate is usually expressed in terms of the ratio of erroneous bits to received bits (eg, 1 in 2×10^4 bits).

Bit rate
The rate at which bits are transmitted expressed in *bits per second* (bps).

Block
A contiguous sequence of data characters transmitted as one unit. Additional characters or codes may be added to the block to permit flow control (eg, synchronisation and error detection).

Block check character
A character tagged to the end of a block which provides a means of verifying that the block has been received without error. The character is derived from a predefined algorithm.

Blocking
In the context of *PBX*, blocking refers to an inability to provide a connection path. In the context of the *data link layer* of the ISO

model for OSI, blocking refers to the combination of serial blocks into one frame.

Bluetooth
A short range radio transmission system used to provide wireless connections to computer peripherals.

Break
A request to terminate transmission.

Broadband
A range of frequencies which is sufficiently wide to accommodate one (or more) carriers modulated by digital information, typically several tens of kilohertz to several tens of megahertz.

Broadband integrated services digital network
An integrated services digital network (ISDN) that is designed to carry digital data, voice and video (see also *integrated services digital network*). *Asynchronous transfer mode* is used to provide packet switching in conjunction with optical fibres and associated high-speed data transmission equipment.

Broadband LAN
A local area network which employs broadband transmission techniques.

Broadband transmission
Transmission method in which a carrier is modulated by a signal prior to being passed through the transmission medium (eg, coaxial cable). Broadband transmission allows several signals to be present within a single transmission medium using *frequency division multiplexing*.

Buffer
In a *hardware* context, a buffer is a device which provides a degree of electrical isolation at an interface (the input to a buffer usually exhibits a much higher impedance than its output). In a *software* context, a buffer is a reserved area of memory which provides temporary data storage and thus may be used to compensate for a difference in the rate of data flow or time of occurrence of events.

Burst errors
A form of error in which several consecutive bits within the transmitted signal are erroneous.

Bus
A signal path which is invariably shared by a number of signals.

Byte
A group of binary digits (*bits*) which is operated on as a unit. A byte normally comprises eight bits and thus can be used to represent a *character*.

Cable
A transmission medium in which signals are passed along electrical conductors (often coaxial).

Carrier
A signal (usually sinusoidal) upon which information is modulated.

Carrier sense
The ability of a node to detect traffic present within a channel.

Carrier sense multiple access
A protocol method which involves listening on a channel before sending. This technique allows a number of nodes to share a common transmission channel.

Central office
A telephone exchange for switching circuits.

Channel
A path between two or more points which allows data communications to take place. Channels are often derived by multiplexing and there need not be a one-to-one correspondence between *channels* and physical *circuits*.

Character
A single letter, figure, punctuation symbol, or control code. Usually represented by either seven or eight bits.

Checksum
A form of error checking in which the sum of all data bytes within a block is formed (any carry generated is usually discarded) and then appended to the transmitted block.

Circuit
An electrical connection comprising two (a *two-wire circuit*) or four wires (a *four-wire circuit*).

Circuit switching
A conventional form of switched interconnection in which a two-way circuit is provided for exclusive use during the period of connection.

Clear
Act of closing a connection.

Clock
A source of timing or synchronising signals.

Close
Act of terminating a connection.

Coaxial cable
A form of cable in which two concentric conductors are employed. The inner conductor is completely surrounded by (but electrically insulated from) the outer conductor. Coaxial cable is commonly used for both *baseband and broadband* LANS.

Collision
A conflict within the transmission path which is caused by two or more nodes sending information at the same time.

Collision avoidance
A technique used to avoid contention in which devices check to see that a network is free before transmitting data.

Collision detection
The process whereby a transmitting node is able to sense a collision.

Common carrier
A national organisation which provides public telecommunications services.

Compression
A technique for reducing the amount of data, whilst not losing any information.

Concentrator
A device which is used to allocate a channel to a number of users on an intelligent time division basis (see also *multiplexer*).

Congestion control
A means of reducing excessive traffic in a network.

Connection
A logical and/or physical relationship between the two end-points of a data link.

Contention
A state which exists when two (or more) users attempt to gain control of a communication channel.

Control character
One, or more, additional characters used to control or facilitate data transmission. Such characters may be responsible for synchronisation, error checking, framing, or delimiting.

Cookie
A file used to store data about the computer and web sites visited.

Cryptography
Security protection by means of encrypted codes.

Current loop
A method of data transmission in which a *mark* (or *logical 1*) is represented by a current in the line while a *space* (or *logical 0*) is represented by the absence of current.

Cyclic redundancy check
An error checking method in which a check character is generated by taking the remainder, after dividing all of the bits within a block of data by a predetermined binary number.

Data
General term used to describe numbers, letters and symbols. The term also encompasses voice, text, fax and video encoded in digital form.

Data access arrangement
Apparatus which allows data communications equipment to be connected to a common carrier network.

Data bit
An individual binary digit (*bit*) which forms part of a serial bit stream in a communications system.

Data communications equipment
Equipment which provides functions that can be used to establish, maintain and terminate a data transmission connection (see also *data terminal equipment*).

Data link layer
A layer within the ISO model for OSI which is responsible for flow control, error detection and link management.

Data set
(see *modem*).

Data terminal equipment
Equipment which is the ultimate source or destination of data (ie, a host computer or microcomputer or a terminal).

Database

An organised collection of data present within a computer storage device. The structure of a database is usually governed by the particular application concerned.

Deadlock

State which occurs when two participating nodes are each waiting for the other to generate a message or acknowledgement and consequently no data transfer takes place.

Demodulation

A process in which the original signal is recovered from a modulated carrier the reverse of modulation. In data transmission, this process involves converting a received analog signal (ie, the modulated carrier) into a baseband digital signal.

Destination node

A node within a network to which a particular message is addressed.

Dial-up method

A method of communication in which a temporary connection is established between two communicating nodes. The connection is terminated when information exchange has been successfully completed.

Dibit encoding

Encoding method in which two bits are handled at a time. In *differential phase shift keying*, for example, each dibit is encoded as one of four unique carrier phase shifts (the four states for a dibit are; 00, 01, 10, and 11).

Differential modulation

A modulation technique in which the coding options relate to a change in some defined parameter of the previously received signal (eg, phase angle).

Digital loopback

A method of testing an item of data communications equipment in which outgoing digital data (*transmitted data*) is connected back to the input of the device (*received data*) and a comparison made (see also *analog loopback*).

Digital signal

A signal that employs only discrete levels of amplitude (see also *analog signal*, and *digital transmission*).

Digital transmission
A method of transmission which employs discrete signal levels (or *pulses*). In practice, two states known variously (and often interchangeably) as *high/low*, *on/off*, *1/10*, and *mark/space*.

Dumb terminal
A terminal which, although it may incorporate local processing and display intelligent functions, is limited in terms of communication protocols.

Duplex
Method of transmission in which information may be passed in both directions (see *full duplex* and *half duplex*).

Echo signal
Distortion that arises when a transmitted signal is reflected (echoed back) to the originating data communications equipment.

Electromagnetic interference
Leakage outside the transmission medium that can cause interference to other services. Cables can be shielded and routed appropriately to reduce the effects of such radiation.

Encryption
A means of rendering data unreadable to unauthorised users.

Equalisation
A technique used to improve the quality of a circuit by minimising distortion.

Error
A condition which results when a received bit within a message is not the same state as that which was transmitted. Errors generally result from noise and distortion present in the transmission path.

Error control
An arrangement, circuit or device which detects the presence of errors and which may, in some circumstances, take steps to correct the errors or request retransmission.

Error rate
The probability, within a specified number of bits, characters, or blocks, of one bit being in error.

Extended binary coded decimal interchange code
A code in which characters are represented as groups of eight bits and which is used primarily in IBM equipment.

File transfer protocol
A protocol used to send file-structured information from one host to another.

Firmware
A program (software) stored permanently in a programmable read-only memory (PROM or ROM) or semi-permanently in erasable-programmable read-only memory (EPROM).

Flag
A symbol having a special significance within a bit-oriented link protocol.

Flow control
A means of controlling data transfer in order to match processing capabilities and/or the extent of buffer storage available.

Fragmentation
Process of dividing a message into pieces or blocks.

Frame
A unit of information at the link protocol level.

Frame check sequence
The error checking information for a frame (eg, a CRC).

Frequency division multiplexing
Transmission technique in which a channel is shared by dividing the available bandwidth into segments occupied by different signals (ie, frequency slicing).

Frequency modulation
A modulation method in which the frequency of a carrier is modified in accordance with the transmitted information.

Frequency shift keying
Technique of modulating digital information onto a carrier by varying its frequency. A logic 1 bit state corresponds to one frequency while a logic 0 bit state corresponds to another frequency.

Front-end processor
A dedicated processor used in conjunction with a larger computer system which handles protocol control, message handling, code conversion, error control, and other specialised functions.

Full duplex
Method of transmission in which information may be passed simultaneously in both directions.

Gateway
A specialised node within a network which provides a means of inter-connecting networks from different vendors.

Half duplex
Method of transmission in which information is passed in one direction at a time.

Handshake
An interlocked sequence of signals between interconnected devices in which each device waits for an *acknowledgment* of its previous signal before proceeding.

Header
The part of a message which contains control information.

Hierarchical network
A network structure in which control is allocated at different levels according to the status of a node.

High-level data link control
The link layer protocol employed in the ISO model and which employs a frame and bit structure as opposed to character-oriented protocols.

High state
The more positive of the two voltage levels used to represent binary logic states. In conventional TTL logic systems, a high state (logic 1) is generally represented by a voltage in the range 2.0 V to 5.0 V.

Host computer
A central computer within a data communications system which provides the primary data processing functions such as computation, database access, etc.

Host–host protocol
End-to-end (transport) protocol.

Impedance
The combined effect of resistance and reactance (either inductive or capacitive) presented by a circuit or device. Like resistance, impedance is measured in ohms. Unlike resistance, the impedance of a circuit or device may be liable to considerable variation with frequency.

Inband control
A transmission technique in which control information is sent over the same channel as the data.

Inband signalling
A signalling technique in which the signalling uses frequencies within the information band of a channel.

Information bit
A bit within a serial bit stream which constitutes part of the transmitted data (ie, not used for flow control or error checking).

Information frame
A frame or bit sequence which contains data.

Input/output (I/O) port
A circuit or functional module that allows signals to be exchanged between a microcomputer system and peripheral devices.

Integrated services digital network
A carrier provided digital service that allows digital data and voice to be accommodated simultaneously (see also *broadband integrated services digital network*).

Interface
A shared boundary between two or more systems, or between two or more elements within a system.

Interface system
The functional elements required for unambiguous communications between two or more devices. Typical elements include: driver and receiver circuitry, signal line descriptions, timing and control conventions, communication protocols, and functional logic circuits.

Internet address
A hardware-independent address assigned to hosts using the TCP/IP protocol. IP version 4 uses a 32-bit address, but IP version 6 uses a 64-bit address.

Internetworking
Communication between two or more networks (which may be of different types).

Isochronous
Transmission method in which all signals are of equal duration and sent in a continuous sequence.

Leased line
A communication line which provides a permanent connection between two nodes. Such a line is invariably leased from a telephone company.

Line driver
A circuit or device which facilitates the connection of a DTE to a line and which handles any necessary level-shifting and electrical buffering in the output (*transmitted data*) path.

Line receiver
A circuit or device which facilitates the connection of a line to a DTE and which handles any necessary level-shifting and electrical buffering in the input (*received data*) path.

Line turnaround
The reversing of transmission direction from sender to receiver and vice versa when using a *half-duplex* circuit.

Link
A channel established between two nodes within a communication system.

Listen-before-talking
A system in which *carrier sense* is employed.

Listen-while-talking
A system in which *collision detection* is employed.

Loaded line
A line to which additional series inductance has been added in order to minimise amplitude distortion. This technique is widely used on public telephone lines in order to improve voice quality. Unfortunately, the presence of appreciable series inductance has the effect of seriously limiting the signalling rate of modems and other data communications equipment that might otherwise be connected to such a line.

Local area network
A network which covers a limited area and which generally provides a high data rate capability. A LAN is invariably confined to a single site (ie, a building or group of buildings).

Local loop
A line which links a subscriber's equipment to a local exchange.

Longitudinal redundancy check
An error detection scheme in which the check character consists of bits calculated on the basis of odd or even parity on all of the characters within the block. Each bit within the longitudinal redundancy check represents a parity bit generated by considering all of the bits within the block at the same position (ie, the first bit of the LRC reflects the state of all of the first bits within the block).

Loopback

A diagnostic test that can be applied to part of a data communications system in which the transmitted signal is returned to the originating device after passing through all or part of the communications link or network (see also *analog loopback* and *digital loopback*).

Low state

The more negative of the two voltage levels used to represent the binary logic states. In a conventional TTL system, a low state (logic 0) is generally represented by a voltage in the range 0 V to 0.8 V.

Mark

A logical 1 or 'on' state (see also *space*).

Memory

Ability of a system to store information for later retrieval.

Message switching

A term used to describe a communication system in which the participants need not he simultaneously connected together and in which data transfer takes place by message forwarding using *store and foreward* techniques.

Microwave link

A communication channel which employs microwave transmission.

Modem

A contraction of modulator–demodulator, a device which facilitates data communication via a conventional telephone line by converting a serial data bit stream into audible signals suitable for transmission over a voice frequency telephone circuit.

Modulation

Technique used for converting digital information into signals which can be passed through an analog communications channel.

Multidrop link

A single line which is shared by a number of nodes. Such links often employ a master or primary node.

Multiple access

A technique which relies upon nodes sensing that a channel is free before sending messages.

Multiplexer

A device which permits *multiplexing* (see also *concentrator*).

Multiplexing
Means by which a communications channel may be shared by several users. *Time division multiplexing* allows users to share a common channel by allocating segments of time to each. *Frequency division multiplexing* allows users to share a common channel by allocating a number of non-overlapping frequency *bands* (*sub-channels*) to users.

Multipoint link
(see *multidrop link*).

Network
A system which allows two or more computers or intelligent devices to be linked via a physical communications medium (eg, coaxial cable) in order to exchange information and share resources.

Network file server
The set of protocols that allow multiple hosts to access files transparently from one another.

Network layer
The layer within the ISO model for OSI which is responsible for services across a network.

Network management system
Equipment, rules and strategies used to monitor, control and manage a data communications network.

Node
An intelligent device (eg, a computer or microcomputer) present within a network. Nodes may be classified as *general-purpose* (eg, a microcomputer *host*) or may have some *network specific* function (eg, *file server*).

Noise
Any unwanted signal component which may become superimposed on a wanted signal. Various types of noise may be present; *Gaussian noise* (or *white noise*) is the random noise caused by the movement of electrons while *impulse noise* (or *black noise*) is the name given to bursts of noise (usually of very short duration) which may corrupt data.

Null modem
A device (usually passive) which allows devices (each configured as a DTE) to exchange data with one another.

Octet
An eight-bit data unit.

Open data link interface
A standard developed by Novell that enables PC adaptor cards to run multiple protocol stacks.

Open systems interconnection
A means of interconnecting systems of different types and from different manufacturers. The ISO model for open systems interconnection comprises seven layers of protocol.

Operating system
A control program which provides a low-level interface with the hardware of a microcomputer system. The operating system thus frees the programmer from the need to produce hardware specific I/O routines (eg, those associated with configuring serial I/O ports).

Optical fibre
A glass or polymer fibre along which signals are propagated optically.

Out of band control
A transmission technique in which control information is sent over a different channel from that occupied by the data.

Pacing
A form of flow control used in systems network architecture, SNA.

Packet
A group of bits (comprising information and control bits arranged in a defined format) which constitutes a composite whole or *unit of information*.

Packet assembler/disassembler
A device which converts asynchronous characters into packets and vice versa.

Packet switched data network
A *vendor-managed* network which employs X.25 protocol to transport data between users' computers. PSDN tariffs are invariably based on the volume of data sent rather than on the distance or connect time.

Packet switching
The technique used for switching within a packet switched data network in which a channel is only occupied for the duration of transmission of a packet. Packets from different users are interleaved and each is directed to its own particular destination.

Parallel transmission
Method of transmission in which all of the bits which make up a character are transmitted simultaneously.

Parity bit

A bit added to an asynchronously transmitted data word which is used for simple error detection (*parity checking*).

Parity check

A simple error checking facility which employs a single bit. Parity may be either *even* or *odd*. The parity bit may be set to logic 1 or logic 0 to ensure that the total number of logic 1 bits present is even (*even parity*) or odd (*odd parity*). Conventionally, odd parity is used in synchronous systems while even parity is employed in asynchronous systems.

Peer entity

A node which has equal status within a network (ie, a logical equal).

Peripheral

An external hardware device whose activity is under the control of a computer or microcomputer system.

Phase modulation

A modulation method in which the phase of a carrier is modified in accordance with the transmitted information.

Physical layer

The lowest layer of the ISO model and which is concerned with the physical transmission medium, types of connector, pin connections, etc.

Piggy-back

A technique for data exchange in which *acknowledgments* are carried with messages.

Pipelining

Technique by which several messages may be in passage at any one time.

Pixel

The smallest element of a computer display. The number of pixels displayed determines the resolution.

Point-to-point link

A network configuration in which one note is connected directly to another.

Polling

Link control by a master slave relation. The master station (eg, a computer) sends a message to each slave (eg, a terminal) in turn to ascertain whether the slave is requesting data.

Port
(see *input/output port*).

Presentation layer
The layer within the ISO model for OSI which resolves the differences in representation of information.

Private line (see *leased line*).

Propagation delay
The time taken for a signal to travel from one point to another.

Protocol
A set of rules and formats necessary for the effective exchange of information within a data communication system.

Pulse code modulation
A modulation method in which analog signals are digitally encoded (according to approximate voltage levels) for transmission in digital form.

Qualified data
A flag (X.25) which indicates how the data packet is to be interpreted.

Query
A request for service.

Queue
A series of messages waiting for onward transmission.

Receiver
Eventual destination for the data within a data transfer.

Redundancy check
A technique used for error detection in which additional bits are added such that it is possible for the receiver to detect the presence of an error in the received data.

Remote procedure calls
A set of functions that allow applications to communicate with a server. Variables and return values are required to support a *client–server architecture*.

Repeater
A signal regenerator.

Residual error rate
The error rate after error control processes have been applied.

Reverse channel
A channel which conveys data in the opposite direction.

Ring network
A network (usually a LAN) which has a circular topology.

Router
A specialised node that enables communication between nodes within a LAN and an X.25 packet switched digital network (see also *gateway*).

Routing
The process of finding a nearly optimal path across a network. An intermediary node (ie, one which is neither a *source node* nor a *destination node*) is often required to have a capability that will facilitate effective routing.

Scroll mode terminal
A terminal in which the data is accepted and displayed on a line-by-line basis.

Sender
The source of data within a data transfer (see *transmitter*).

Serial transmission
Method of transmission in which one bit is transmitted after another until all of the bits which represent a character have been sent.

Session layer
The layer in the ISO model which supports the establishment, control and termination of dialogues between application processes.

Sideband
The upper and lower frequency bands which contain modulated information on either side of a carrier and which are produced as a result of modulation.

Signal
Information conveyed by an electrical quantity.

Signal level
The relative magnitude of a signal when considered in relation to an arbitrary reference (usually expressed in volts).

Signal parameter
That element of an electrical quantity whose values or sequence of values is used to convey information.

Simplex
Method of transmission in which information may be passed in one direction only.

Sliding window
A mechanism which indicates the frame or frames that can currently he sent.

Socket
An entry and/or exit point (see also *input/output port*).

Source node
A node within a network which is the originator of a particular message.

Source routing
A process which determines the path or route of data at the source of the message.

Space
A logical 0 or 'off' state (see also *mark*).

Start bit
The first bit (normally a *space*) of an asynchronously transmitted data word which alerts the receiving equipment to the arrival of a character.

Start/stop signalling
Asynchronous transmission of character.

Statistical multiplexer
(see *concentrator*).

Stop and wait protocol
A protocol which involves waiting for an *acknowledgment* (eg, ACK) before sending another message.

Stop bits
The last bit (or bits), normally *mark*, of an asynchronously transmitted data word which signals that the line is about to be placed in its rest state.

Store and forward
A process in which a message or *packet* is stored temporarily before onward transmission.

Supervisory frame
A *control frame*.

Switching
A means of conveying information from source to destination across a network.

Synchronisation
Establishing known timing relationships.

Synchronous data link control
IBM standard communication protocol which replaces *binary synchronous communications*.

Synchronous transmission
Method of transmission in which data is transmitted at a fixed rate and in which the transmitter and receiver are both synchronised.

Tandem
A network configuration in which two or more point-to-point circuits are linked together with transmission effected on an end-to-end basis over all links.

Terminal server
A special-purpose node which allows a number of terminals to he connected to a network via a single physical line. A terminal server thus frees network nodes from the burden of establishing connections between local terminals and remote nodes. Terminals connected to a terminal server will, of course, have access to all nodes present within the network.

Time division multiplexing
Transmission technique in which users share a common channel by allocating segments of time to each (ie, *time slicing*).

Time sharing
A method of operation in which a computer facility is shared by a number of users. The computer divides its processing time between the users and a high speed of processing ensures that each user is unaware of the demands of others and processing appears to be virtually instantaneous.

Timeout
Period during which a predetermined time interval has to elapse before further action is taken (usually as a result of no response from another node).

Token
A recognisable control mechanism used to control access to a network.

Topology
The structure of a network and which is usually described in the form of a diagram which shows the nodes and links between them.

Traffic analysis
Process of determining the flow and volume of traffic within a network.

Transceiver
A transmitter/receiver.

Transmitter
Source of data (see *sender*).

Transparency
A property of a network that allows users to access and transfer information without being aware of the physical, electrical and logical characteristics of the network.

Transport layer
The layer of the ISO model for OSI which describes host–host communication.

Tribit encoding
Encoding method in which three bits are handled at a time.

Trunk
A single circuit between two switching centres or distribution points. Trunks normally provide a large number of channels of communication simultaneously.

Unbalanced line
A transmission line in which a single conductor is used to convey the signal in conjunction with a ground or earth return. A coaxial cable is an example of an unbalanced line (see also *balanced line*).

Unnumbered frame
A control frame.

V-series
A series of recommendations specified by the CCITT which defines analog interface and modem standards for data communications over common carrier lines such as a PSDN.

Vertical redundancy check
An error detection scheme in which one bit of each data word (the *parity bit*) is set to logic 1 or logic 0 so that the total number of logic 1 bits is odd (*odd parity*) or even (*even parity*).

Video on demand
A pay per view television service, often provided over ADSL line equipment.

Virtual circuit
An arrangement which provides a sequenced, error-free delivery of data.

Voice-grade line
A conventional telephone connection.

Wide area network
A network which covers a relatively large geographical area (eg, one which spans a large region, state, country or continent).

Wideband
A communications channel which exhibits a very much greater bandwidth than that associated with a conventional voice-grade channel and which will support data rates of typically between 10k and 500 kbps.

Workstation
A general-purpose node within a network which provides users with processing power, and which is invariably based on a PC or other microcomputer.

X-series
A series of recommendations specified by the CCITT which defines digital data communications over common carrier lines such as a PSDN.

Zero insertion
Transparency method for bit-orientated link protocols.

Abbreviations used for advisory bodies and other organisations

ACTs	advisory committees on telecommunications
ANSI	American National Standards Institute
ARPA	Advanced Research Projects Agency
ASA	American Standards Association
AT&T	American Telephone and Telegraph Corporation
BABT	British Approvals Board for Telecommunications
BEITA	Business Equipment and Information Technology Trade Association

BFIC	British Facsimile Industry Consultative Committee
BREEMA	British Radio and Electronic Equipment Manufacturers' Association
BSI	British Standards Institution
BT	British Telecom
CCITT	International Telephone and Telegraph Consultative Committee (now ITU-T)
CEPT	European Conference of Postal and Telecommunications Administrations
COMSAT	Communications Satellite Corporation
CSA	Canadian Standards Association
DTI	Department of Trade and Industry
EARN	European Academic Research Network
ECMA	European Computer Manufacturer's Association
EEA	Electronic Engineering Association
EIA	Electronics Industries Association
ETSI	European Telecommunications Standards Institute
FCC	Federal Communications Commission
IBM	International Business Machines
IEE	Institution of Electrical Engineers
IEEE	Institution of Electrical and Electronic Engineers
IEEIE	Institution of Electrical and Electronics Incorporated Engineers
IERE	Institution of Electronic and Radio Engineers
IETF	Internet Engineering Task Force
INTELSAT	International Telecommunications Satellite Consortium
ISO	International Standards Organisation
ITU	International Telecommunication Union
NBS	National Bureau of Standards
NCC	National Computing Centre
NIST	National Institute of Standards and Technology
PATACS	Posts and Telecommunications Advisory Committee
PTT	Postal, Telegraph and Telephone authority
SITA	Société Internationale de Telecommunication Aeronautique
SWIFT	Society for Worldwide Interbank Financial Telecommunications
TEMA	Telecommunication Engineering and Manufacturing Association
TMA	Telecommunications Managers Association

2 Terminals

Terminals are used to enter data into computer systems and, as such, can be considered a data source. The older style of keyboard is the teletype that was, and in some cases still is, used to enter data remote from a mainframe computer. Data is transmitted over telex circuits or a radio channel using IA2 and IA5 coded signals. However, use of this form of coding is now quite rare. This chapter includes details of IA2, IA5 and EBCDIC.

One of the more common terminals is VT-100. Modern personal computers can be used to emulate VT-100 terminals (e.g. using the HyperTerminal program that comes with Windows™). Details of VT-52, VT-100 and WYSE 100 terminal control codes are included here.

Representative teletype keyboard layout (IA2)

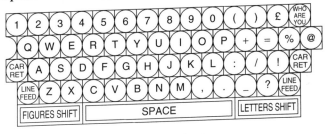

International alphabet no. 2 (IA2)

Position	Letters shift enabled	Figures shift enabled
12345		
●●	A	–
● ●●	B	?
●●●	C	:
● ●	D	who are you?
●	E	3
● ●●	F	
● ●●	G	
● ●	H	
●●	I	8
●● ●	J	Bell
●●●●	K	(
● ●	L	)
●●●	M	.
●●	N	,
●●	O	9
●● ●	P	0
●●● ●	Q	1
● ●	R	4
● ●	S	!
●	T	5
●●●	U	7
●●●●	V	=
●● ●	W	2
● ●●●	X	/
● ● ●	Y	6
● ●	Z	+
●●●●●	Blank	
●● ●●	Letters shift	
●	Figure shift	
● ●	Space	
●	Carriage return	
	Line feed	

Notes:

1. ● = punched holes in paper tape media
2. Sprocket feed holes are located between positions 2 and 3 on paper tape media

International alphabet no. 5 (IA5)

IA5 standard code table

$b_3b_2b_1b_0$	row	$b_6b_5b_4$ col	000	001	010	011	100	101	110	111
			0	1	2	3	4	5	6	7
0000	0		NUL	DLE	SP	0	@	P		p
0001	1		SOH	DC1	!	1	A	Q	a	q
0010	2		STX	DC2	"	2	B	R	b	r
0011	3		ETX	DC3	£/#	3	C	S	c	s
0100	4		EOT	DC4	$	4	D	T	d	t
0101	5		ENQ	NAK	%	5	E	U	e	u
0110	6		ACK	SYN	&	6	F	V	f	v
0111	7		BEL	ETB	/	7	G	W	g	w
1000	8		BS	CAN	(	8	H	X	h	x
1001	9		HT	EM	)	9	I	Y	i	y
1010	10		LF	SUB	*	:	J	Z	j	z
1011	11		VT	ESC	+	;	K	[	k	{
1100	12		FF	FS	,	<	L	\	l	:
1101	13		CR	GS	–	=	M	]	m	}
1110	14		SO	RS	.	>	N	^	n	"
0111	15		SI	US	/	?	O	_	o	DEL

(The leftmost two header columns are labelled "Most significant bits" spanning columns 000–111; the left side is labelled "Least significant bits".)

IA5 control characters

Character	Full name	Function
(a) Logical communication control		
ACK	acknowledge	indicates an affirmative response (transmitted by a receiver to acknowledge that data has been received without error)
DLE	data link escape	marks the start of a contiguous sequence of characters which provide supplementary data transmission control functions (only graphics and transmission control characters appear in DLE sequences)
ENQ	enquiry	requests a response from a remote station which may either take the form of a station identification or status (the first use of ENQ after a connection has been established is equivalent to 'who are you?')
EOT	end of transmission	concludes the transmission (and may also terminate communications by turning a device off)
ETB	end of transmission block	transmission block (unrelated to any division in the format of the logical data itself)

Character	Full name	Function
ETX	end of text	indicates the last character in the transmission of text (often generated by means of CTRL-C in many terminals)
NAK	negative acknowledge	indicates a negative response (the opposite of ACK)
SOH	start of heading	indicates the first character of the heading of an information message
STX	start of text	terminates a heading and indicates that text follows
SYN	synchronous idle	provides a signal which may be needed to achieve (or retain) synchronisation between devices (used in the idle condition when no other characters are transmitted)

(b) Physical communication control

CAN	cancel	indicates that the preceding data is to be disregarded (it may contain errors). CAN is usually employed on a line-by-line basis such that, when CAN appears within a serial data stream, data is disregarded up to the last CR character received. On most terminals, CAN is generated by CTRL-X
DEL	delete	DEL was originally used to obliterate unwanted characters in punched tape. However, in applications where it will not affect the information content of a data stream, DEL may be used for media or time-fill (see note 1)
EM	end of medium	identifies the end of the used portion of the medium (not necessarily the physical end of the medium)
NUL	null	may be inserted into, or removed from, the data stream without affecting the information content (and may thus be used to accomplish media or time-fill)
SUB	substitute	may be used to replace a suspect character (ie one which is for one reason or another considered invalid)

(c) Device control

BEL	bell	produces an audible signal to attract the user's attention
BS	backspace	a layout character which moves the printing position backwards by one character print position (often generated by CTRL-H). With hardcopy devices, BS can be used for a variety of purposes including underlining, bold highlighting, and the generation of composite characters
CR	carriage return	a layout character which moves the printing position to the start of the *current* line

Character	Full name	Function
DC1–DC4	device control	used to enable or disable additional facilities which may be available at the receiver (often used to control specialised printing functions)
FF	form feed	a layout character which moves the printing position to the first printing line on the next page (form)
HT	horizontal tabulation	a layout character which moves the printing position to the next in a series of predefined horizontal printing positions (horizontal tab settings)
LF	line feed	a layout character which moves the printing position to the next printing line. In some equipment, LF is sometimes combined with CR so that the print position is moved to the start of the *next* line. To avoid confusion, LF is sometimes referred to as NL (or new line)
VT	vertical tabulation	a layout character which moves the printing position to the next in a series of predefined vertical printing positions (vertical tab. settings). Depending upon the current vertical tab. setting, VT is equivalent to one, or more, LF characters.

(d) Formatting and string processing (see note 2)

FS	field separator	terminates a file information block
GS	group separator	terminates a group information block
RS	record separator	terminates a record information block
US	unit separator	terminates a unit information block

(e) Character/graphic set control

ESC	escape	used to modify or extend the standard character set. The escape character changes the meaning of the character which follows according to some previously defined scheme. NUL, DEL communication control characters must not be used in defining escape sequences
SI	shift-in	characters which follow SI should be interpreted according to the standard code table
SO	shift-out	characters which follow SO should be interpreted as being outside the standard code table. The meaning of the control characters from columns 0 and 1 are, however, preserved.

Notes:

1. Note that DEL, unlike other control characters which occupy columns 0 and 1, is in column 7 (all bits of the code for DEL are set to logic 1)
2. Information block separators have the following hierarchy (arranged in ascending order): US, RS, GS, FS. Also note that information blocks may not themselves be divided by separators of higher order

Extended binary coded decimal interchange code (EBCDIC)

EBCDIC standard code table

Most significant bits

b3b2b1b0 \\ b7b6b5b4	0000	0001	0010	0011	0100	0101	0110	0111	1000	1001	1010	1011	1100	1101	1110	1111
row \\ col	**0**	**1**	**2**	**3**	**4**	**5**	**6**	**7**	**8**	**9**	**10**	**11**	**12**	**13**	**14**	**15**
0000	NULL				SP	&	–									0
0001							/		a	j			A	J		1
0010									b	k	s		B	K	S	2
0011									c	l	t		C	L	T	3
0100	PF	RES	BYP	PN	¢	!			d	m	u		D	M	U	4
0101	HT	NL	LF	RS	.	$			e	n	v		E	N	V	5
0110	LC	BS	EOB	UC	<	*			f	o	w		F	O	W	6
0111	DEL	IL	PRE	EOT	(	)			g	p	x		G	P	X	7
1000					+	;			h	q	y		H	Q	Y	8
1001									i	r	z		I	R	Z	9
1010			SM		¢	!	¦	:								
1011					.	$	,	#								
1100					<	*	%	@								
1101					(	)	_	'								
1110					+	;	>	=								
1111					|	¬	?	"								

Least significant bits

EBCDIC control characters

Character	Full name
BS	Backspace
BYP	Bypass
DEL	Delete
EOB	End of block
EOT	End of transmission
HT	Horizontal tab
IL	Idle
LC	Lower case
LF	Line feed
NL	New line
NULL	Null/idle
PF	Punch off
PN	Punch on
PRE	Prefix
RES	Restore
RS	Reader stop
SM	Set mode
SP	Space
UC	Upper case

Representative personal computer keyboard layout

Terminal control codes

Control code sequences are used by terminals to provide special functions such as deletion of the character at the cursor position, clearing the entire screen display, and horizontal tabulation. In addition, special function keys or keypads may be provided and these also produce control code sequences (often beginning with the ASCII ESCape character (1B hex). The following is a list of the control code sequences used in some of the most popular computer terminals:

VT-52

Key	Control code sequence (hexadecimal)
Horizontal tab	09
Character delete	7F
Home cursor	1B,48
Cursor up	1B,41
Cursor down	1B,42
Cursor left	1B,44
Cursor right	1B,43
Clear screen	1B,48,1B,4A
Erase end of line	1B,4B
Keypad application mode 0	1B,3F,70
Keypad application mode 1	1B,3F,71
Keypad application mode 2	1B,3F,72
Keypad application mode 3	1B,3F,73
Keypad application mode 4	1B,3F,74
Keypad application mode 5	1B,3F,75
Keypad application mode 6	1B,3F,76
Keypad application mode 7	1B,3F,77
Keypad application mode 8	1B,3F,78
Keypad application mode 9	1B,3F,79
Keypad application mode −	1B,3F,6D
Keypad application mode ,	1B,3F,6C
Keypad application mode .	1B,3F,6E
Keypad application mode ENTER	1B,3F,4D
Program function 1 (PF1)	1B,50
Program function 2 (PF2)	1B,51
Program function 3 (PF3)	1B,52
Program function 4 (PF4)	1B,53

VT-100

Key	Control code sequence (hexadecimal)
Horizontal tab	09
Character delete	7F
Home cursor	1B,5B,48
Cursor up	1B,5B,41
Cursor down	1B,5B,42
Cursor left	1B,5B,44
Cursor right	1B,5B,43
Clear screen	1B,5B,48,1B,5B,32,4A
Erase end of line	1B,5B,4B
Insert line	1B,5B,4C

Key	Control code sequence (hexadecimal)
Delete line	1B,5B,4D
Line feed	0A
Keypad application mode 0	1B,4F,70
Keypad application mode 1	1B,4F,71
Keypad application mode 2	1B,4F,72
Keypad application mode 3	1B,4F,73
Keypad application mode 4	1B,4F,74
Keypad application mode 5	1B,4F,75
Keypad application mode 6	1B,4F,76
Keypad application mode 7	1B,4F,77
Keypad application mode 8	1B,4F,78
Keypad application mode 9	1B,4F,79
Keypad application mode −	1B,4F,6D
Keypad application mode ,	1B,4F,6C
Keypad application mode .	1B,4F,6E
Keypad application mode ENTER	1B,4F,4D
Program function 1 (PF1)	1B,4F,50
Program function 2 (PF2)	1B,4F,51
Program function 3 (PF3)	1B,4F,52
Program function 4 (PF4)	1B,4F,53

WYSE 100

Key	Control code sequence (hexadecimal)
Horizontal tab	09
Reverse tab	1B,49
Insert character	1B,51
Insert line	1B,45
Delete character	7F
Delete line	1B,52
Home cursor	1E
Cursor up	1B
Cursor down	0A
Cursor left	18
Cursor right	1C
Clear screen	1A
Line erase	1B,54
Page erase	1B,59
Function 1 (F1)	01,40,0D
Function 2 (F2)	01,41,0D
Function 3 (F3)	01,42,0D

Key	Control code sequence (hexadecimal)
Function 4 (F4)	01,43,0D
Function 5 (F5)	01,44,0D
Function 6 (F6)	01,45,0D
Function 7 (F7)	01,46,0D
Function 8 (F8)	01,47,0D
Shift function 1 (F1)	01,48,0D
Shift function 2 (F2)	01,49,0D
Shift function 3 (F3)	01,4A,0D
Shift function 4 (F4)	01,4B,0D
Shift function 5 (F5)	01,4C,0D
Shift function 6 (F6)	01,4D,0D
Shift function 7 (F7)	01,4E,0D
Shift function 8 (F8)	01,4F,0D

Commonly used control characters (with keyboard entry)

Keyboard	Decimal	Binary	Hexa-decimal	ASCII	Function (eg, MS-DOS)
CTRL-@	0	00000000	00	NUL	
CTRL-A	1	00000001	01	SOH	
CTRL-B	2	00000010	02	STX	
CTRL-C	3	00000011	03	ETX	Cancels (if possible) the current process or aborts the current program (ie, same effect as CTRL-BREAK)
CTRL-D	4	00000100	04	EOT	
CTRL-E	5	00000101	05	ENQ	
CTRL-F	6	00000110	06	ACK	
CTRL-G	7	00000111	07	BEL	Bell (not normally executable directly from the keyboard)
CTRL-H	8	00001000	08	BS	Backspace (same as BS or left arrow keys)
CTRL-I	9	00001001	09	HT	Tab (usually eight print positions to the right). Same effect as TAB key
CTRL-J	10	00001010	0A	LF	Line feed. Moves the print position to the next line. Same effect as CTRL-RETURN
CTRL-K	11	00001011	0B	VT	
CTRL-L	12	00001100	0C	FF	Form feed. Moves the print position to the corresponding point on the next page/form

Keyboard	Decimal	Binary	Hexa-decimal	ASCII	Function (eg, MS-DOS)
CTRL-M	13	00001101	0D	CR	Carriage return. Same effect as the RETURN key
CTRL-N	14	00001110	0E	SO	Enables expanded mode printing (EPSON)
CTRL-O	15	00001111	0F	SI	Enables condensed mode printing (EPSON)
CTRL-P	16	00010000	10	DLE	Print. Toggles (on or off) the echoing of characters printed on the screen to a line printer. Same effect as CTRL-PRT SCN
CTRL-Q	17	00010001	11	DC1	X-ON (resumes flow)
CTRL-R	18	00010010	12	DC2	Disables condensed mode printing (EPSON)
CTRL-S	19	00010011	13	DC3	X-OFF (halts flow). May be used to interrupt flow of characters when a TYPE command is being executed
CTRL-T	20	00010100	14	DC4	Disables expanded mode printing (EPSON)
CTRL-U	21	00010101	15	NAK	
CTRL-V	22	00010110	16	SYN	
CTRL-W	23	00010111	17	ETB	
CTRL-X	24	00011000	18	CAN	Cancel text in buffer (EPSON)
CTRL-Y	25	00011001	19	EM	
CTRL-Z	26	00011010	1A	SUB	End-of-file (EOF)
CTRL-[	27	00011011	1B	ESC	Escape (same effect as an ESC key)
CTRL-\	28	00011100	1C	FS	
CTRL-]	29	00011101	1D	GS	
CTRL-ˆ	30	00011110	1E	RS	
CTRL-_	31	00011111	1F	US	
SPACE	32	00100000	20	SP	Generated by the space bar

Notes:

1. CTRL is often represented by the character 'ˆ'. Hence CTRL-A (control-A) may be shown as ˆA
2. When entering control characters from a keyboard, the control key (CTRL) must be held down *before* the other keyboard character is depressed
3. Control characters can usually be incorporated in BASIC programs by using statements of the form: (L)PRINT CHR$(n). To produce condensed mode printing on an EPSON printer, for example, the following BASIC statement is used: LPRINT CHR$(15)

3 Transmission media

Data communications is about moving or copying data from one place to another. This may be from a personal computer to a file server on a local area network (LAN), or may be from the Internet to a personal computer. In all cases data must be carried over a cable at some point; this could be a copper cable or an optical fibre cable.

This chapter describes the types of cable and their performance. It is important to know the limitations of the transmission media, in order to understand why modems, repeaters and other data communications equipment (DCE) is necessary.

Transmission element specifications

The transmission path in a data communications system may comprise cables, amplifiers/regenerators, attenuators, filters, diplexers, etc. The electrical characteristics of such items are usually specified in terms of one or more of the following parameters.

Gain or loss

The gain or loss of an element within a transmission path is the ratio of output voltage to input voltage (ie, voltage gain), output current to input current (ie, current gain), or output power to input power (ie, power gain). Gain is often expressed in decibels (dB) where:

$$\text{voltage gain in dB} = 20 \log_{10} \left(\frac{V_{\text{out}}}{V_{\text{in}}} \right)$$

$$\text{current gain in dB} = 20 \log_{10} \left(\frac{I_{\text{out}}}{I_{\text{in}}} \right)$$

$$\text{power gain in dB} = 10 \log_{10} \left(\frac{P_{\text{out}}}{P_{\text{in}}} \right)$$

Note that in the two former cases, the specification is only meaningful where the input and output impedances of the element are identical.

Input impedance

The input impedance of an element within a transmission path is the ratio of input voltage to input current and it is expressed in ohms. The input of an amplifier is normally purely resistive (ie, the reactive component is negligible) in the middle of its working frequency range (ie, the mid-band) and hence, in such cases, input impedance is synonymous with input resistance.

Output impedance

The output impedance of an element within a transmission path is the ratio of open-circuit output voltage to short-circuit output current and is measured in ohms. Note that this impedance is internal to the element and should not be confused with the impedance of the load or circuit to which the element is connected. (Usually, but not always, these will have identical values in order to maximise power transfer).

Frequency response

The frequency response of a transmission element is usually specified in terms of the upper and lower cut-off frequencies of the element. These frequencies are those at which the output power has dropped to 50% (otherwise known as the −3dB points) or where the voltage gain has dropped to 70.7% of its mid-band value.

Bandwidth

The bandwidth of a transmission element is usually taken as the difference between the two cut-off frequencies. It is equivalent to the frequency span for which the gain is maintained within defined limits (usually within 3dB of the mid-band power gain).

Phase shift

The phase shift of a transmission element is defined as the phase angle (in electrical degrees or radians) of the output signal when compared with the input signal (taken as the reference). Phase shift is substantially constant within the mid-band region but is liable to a marked variation beyond cut-off due to the increasing significance of reactance.

Equivalent circuit of a transmission element

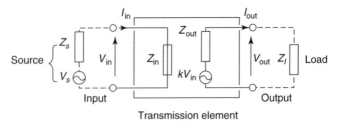

Transmission element

Frequency response of a transmission element

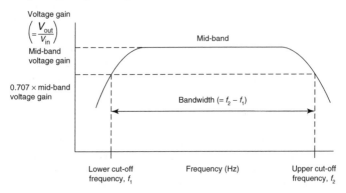

Data cable types

Many different types of cable are employed in data communications ranging from simple twisted-pair to multi-core coaxial. For uncritical applications where speed and distance are both limited, twisted-pair cables are perfectly adequate. However, for more critical applications which involve high data rates and longer distances, high quality low-loss coaxial cables are essential. Furthermore, to minimise the effects of crosstalk, induced noise and radiation, individual and overall braided or foil screens may be required. The following diagrams (courtesy of BICC) are provided in order to assist readers in identifying the major types of cable which are in current use.

Multi-core (unscreened)

Multi-core with overall braid screen

Multi-core with individually screened conductors

Two-pair cable with overall braid screen

Single-pair cable with foil screen

Two-pair cable with overall braid and foil screens (stranded signal conductors)

Two-pair cable with overall braid and foil screens (solid signal conductors)

Multi-pair cable with overall foil screen

Multi-pair cable with individual foil screens

Multi-pair cable with overall braid and foil screens

Coaxial cable with foil and braid screens

Coaxial cable with double braid screen and foil (Ethernet trunk)

Multi-pair with individual foil and overall braid screens (Ethernet transceiver drop)

Two-pair with individual foil and overall braid screens (IBM indoor data cable)

*Four-pair with individual foil and overall foil and braid
screens (DECconnect transceiver cable)*

Four-pair unscreened (DECconnect four-pair cable)

Flat six-way unscreened (DECconnect cordage)

Coaxial cable with braid screen and solid centre conductor

*Dual coaxial cable with individual braid screens and solid
centre conductors*

Coaxial cable with double braid screens

Simplex optical cable

Duplex optical cable

Coaxial cable data

Type	Centre conductor	Diameter (mm)	Impedance (ohm)	Capacitance (pF m^{-1})	Attenuation (dBm^{-1})
RG6/U	1/1.02 mm	68.6	75	56.8	0.069 at 100 MHz
RG11A/U	7/0.41 mm	10.3	75	67	
RG58C/U	19/0.18 mm	4.95	50	100	0.2 at 10 MHz
					0.31 at 200 MHz
					0.76 at 1 GHz
RG59B/U	1/0.58 mm	6.15	75	60.6	0.12 at 100 MHz
					0.19 at 200 MHz
					0.3 at 400 MHz
					0.46 at 1 GHz
RG59/U	1/0.64 mm	6.15	75	56.8	0.098 at 100 MHz
RG62A/U	1/0.64 mm	6.15	93	36	0.26 at 400 MHz
RG174/U	1/0.4 mm	2.56	60	101	0.292 at 100 MHz
RG174A/U	7/0.16 mm	2.54	50	100	0.11 at 10 MHz
					0.42 at 200 MHz
					0.67 at 400 MHz
RG178B/U	7/0.1 mm	1.91	50	106	0.18 at 10 MHz
					0.44 at 100 MHz
					0.95 at 400 MHz
					1.4 at 1 GHz
RG179B/U	7/0.1 mm	2.54	75	66	0.19 at 10 MHz
					0.32 at 100 MHz
					0.69 at 400 MHz
					0.82 at 1 GHz
RG188A/U	7/0.17 mm	2.6	50	93	
RG213/U	7/0.029 mm	10.29	50	98	0.18 at 400 MHz
RG214/U	7/0.029 mm	10.79	50	98	0.18 at 400 MHz
RG223/U	1/0.9 mm	5.5	50	96	
RG316/U	7/0.17 mm	2.6	50	102	
URM43	1/0.9 mm	5	50	100	0.13 at 100 MHz
					0.187 at 200 MHz
					0.232 at 300 MHz
					0.338 at 600 MHz
					0.446 at 1 GHz
URM57	1/1.15 mm	10.3	75	67	0.061 at 100 MHz
					0.09 at 200 MHz
					0.113 at 300 MHz
					0.17 at 600 MHz
					0.231 at 1 GHz
URM67	7/0.77 mm	10.3	50	100	0.068 at 100 MHz
					0.099 at 200 MHz
					0.125 at 300 MHz
					0.186 at 500 MHz
					0.252 at 1 GHz

Type	Centre conductor	Diameter (mm)	Impedance (ohm)	Capacitance $(pF\,m^{-1})$	Attenuation $(dB\,m^{-1})$
URM70	7/0.19 mm	5.8	75	67	0.152 at 100 MHz 0.218 at 200 MHz 0.27 at 300 MHz 0.391 at 600 MHz 0.517 at 1 GHz
URM76	7/0.32 mm	5	50	100	0.155 at 100 MHz 0.222 at 200 MHz 0.274 at 300 MHz 0.398 at 600 MHz 0.527 at 1 GHz
URM90	1/0.6 mm	6	75	67	1.12 at 100 MHz 3.91 at 1 GHz
URM95	1/0.46 mm	2.3	50	100	0.27 at 100 MHz 0.37 at 200 MHz 0.46 at 300 MHz 0.65 at 600 MHz
URM96	1/0.64 mm	6	95	40	0.79 at 100 MHz 2.58 at 1 GHz
URM202	7/0.25 mm	5.1	75	56	0.086 at 60 MHz 0.11 at 100 MHz 0.16 at 200 MHz 0.27 at 500 MHz 0.4 at 900 MHz
URM203	1/1.12 mm	7.25	75	56	0.057 at 60 MHz 0.075 at 100 MHz 0.11 at 200 MHz 0.185 at 500 MHz 0.26 at 900 MHz
2001	7/0.2 mm	4.6	75	56.7	0.04 at 5 MHz 0.14 at 60 MHz 0.253 at 200 MHz
2002	7/0.2 mm	5.2	75	56.7	0.0126 at 1 MHz 0.042 at 10 MHz 0.138 at 100 MHz
2003A	7/0.2 mm	6.9	75	67	0.026 at 5 MHz 0.09 at 60 MHz 0.185 at 200 MHz
Ethernet trunk cable	1/2.17 mm	10.3	50	85	0.02 at 5 MHz 0.04 at 10 MHz

Screened and unscreened pair data

Type	Diameter (mm)	Impedance (ohm)	Capacitance between conductors (pF m⁻¹)	Capacitance between conductor and screen (pF m⁻¹)	Attenuation (dBm⁻¹)
BICC H8071/Belden 9501	4.6	62	135	246	0.062 at 1 MHz 0.15 at 10 MHz
BICC H8072/Belden 9502 (2 pair)	5.6	77	98	164	0.062 at 1 MHz 0.15 at 10 MHz
BICC H8073/Belden 9504 (4 pair)	6.7	77	98	154	0.062 at 1 MHz 0.15 at 10 MHz
BICC H8074/Belden 9506 (6 pair)	7.6	77	98	164	0.062 at 1 MHz 0.15 at 10 MHz
H8082/Belden 8761	5.4	85	79	154	
BICC H8085/Belden 8723 (2 pair)	4.19	54	115	203	
BICC H8086/Belden 8777 (3 pair)	7.9	62	98	180	
BICC H8088/Belden 8774 (9 pair)	11.9	62	98	180	
BICC H8150/Belden 8795	4.0	110	56	n/a	0.016 at 1 MHz 0.063 at 10 MHz
Belden 8205/Alpha 1895	4.8		55	n/a	
Belden 8761	4.6	85	79	154	
Belden 9855/Alpha 9819	7.7	108	46		0.02 at 1 MHz
Belden 9891 (4 pair)	10.03	78	64.6	113.8	
Belden 9892 (4 pair)	10.67	78	64.6	113.8	
Belden 9893 (5 pair)	12.95	78	64.6	113.8	

Note: n/a – not applicable (unscreened cable)

CAT-3, -4, -5, -6, -7 cable

Twisted pair data cables for LANs (such as 10 base T or 100 base T) are described as category 3, 4, 5, 6 or 7; these are often referred to as CAT-3, CAT-4, etc. CAT-5 is specified by standards TIA/EIA 568A, ISO/IEC 11801, EN 50173. CAT-6 is specified by standards TIA/EIA 568B, ISO/IEC 11801 Category 6, EN 50288. The category determines the maximum data rate over 100 metres of cable:

Category	Data rate
CAT-3	10 Mbit/s
CAT-4	20 Mbit/s
CAT-5	100 Mbit/s
CAT-5e/6	350 Mbit/s
CAT-7	1 Gbit/s

Note 1: 10base-T and 100base-TX transmit over two pairs (one transmit and one receive), thus 10base-T requires CAT-3 cable and 100base-TX requires CAT-5 cable. However, 100base-T4 transmits and receives over four pairs, allowing CAT-3 cable to be used. 1000base-T [1 Gbit/s] can be carried over CAT-5e cable by transmitting 250 Mbit/s over each pair.

Note 2: Distances greater than 100 m can be achieved by operating at a lower data rate: ie, CAT-3 cable can be used to transmit 1 Mbit/s over 250 m.

All categories of cable comprise four twisted copper pairs; each pair being two insulated wires of 0.5 mm diameter (24 a.w.g.) solid wire. The wire insulation affects the transmission performance of the cable and typically PVC is used in CAT-3 and CAT-4 cable, but more expensive Polyolefin is used in CAT-6 or CAT-7 cable. In all cases, the impedance of the pair is about 100 ohms.

The standard colour code for the wire insulation material is as follows:

Pair number	Primary colour	Secondary colour	Jack wiring (TIA 568B)
1	blue	white	4,5
2	orange	white	2,1
3	green	white	6,3
4	brown	white	8,7

The copper pairs are enclosed within a plastic sheath, which is typically made from PVC, Polyolefin or low-smoke/fume (LSF) material. Foiled twisted pair (FTP) cables have an overall metal foil layer inside the plastic sheath; a copper drain wire is provided for ease of connection to an earthing point at the cable termination. Screened twisted pairs (STP) have a metal braid inside the plastic sheath. Unshielded twisted pair (UTP) cables are more common because they are easier to handle and terminate.

Cable equivalents

Alpha	Belden	BICC	Brand Rex	Notes
1895	8205			Unscreened pair
	8216	T3390		
	8259	T3429		
2401	8761	H8082	BE-56761	
	8262	T3428		
	8263	T3429		
2461	8451	H8084	BI-56451	
2466	8723	H8085	BI-56723	2-pair 58 ohm, UL2493
2401	8761	H8082		UL2092
2403	8771	H8101		UL2093
6022	8773	H8118	BE-56773	27-pair 55 ohm, UL2919
6014	8774	H8088	BE-56774	9-pair 55 ohm, UL2493
6010	8777	H8086	BE-56777	3-pair 55 ohm, UL2493
1202	8795	H8105		Unscreened-pair 110 ohm
	9204	T3429		
9817	9207	H8106	BC-57207	IBM7362211, UL2498
9818	9207	H8106	BC-57207	IBM7362211, UL2498
9063	9269	T3430		RG62A/U, UL1478
9815	9272	H8065	BC57272	Twin-axial 78 ohm, UL2092
5902	9302	H8079	BE-57302	
5471	9501	H8071		1-pair 62 ohm, UL2464
5472	9502	H8072	BE-57502	2-pair 77 ohm, UL2464
5473	9503	H8136	BE-57503	3-pair 77 ohm, UL2464
5474	9504	H8073	BE-57504	4-pair 77 ohm, UL2464
5475	9505	H8173	BE-57505	5-pair 77 ohm, UL2464
5476	9506	H8074	BE-57506	6-pair 77 ohm, UL2464
5480	9510	H8133	BE-57510	10-pair 85 ohm, UL2464
9845	9555	H8119	BC-57555	RG59B/U, Wang 420-0057
	9696	H8064		
	9729	H9002	BE-57555	UL2493
6017	9768	H8113	BE-57768	12-pair 55 ohm, UL2493
	9829	H9564		UL2919
9819	9855	H8063		UL2919, UL2582
	9880	H8112	BC-57880	Ethernet trunk coaxial
	9881			Multicore + coax, UL2704
	9892		BN-57892	Ethernet, 4-pair
		2002		ICL80047293, UL 1354
		H9601	GT-75340	ICL80049808
			GT-553011	ICL80048808, Oslan
			GT-551014	ICL80049496, Cheapernet

Important note:
Cable types listed above may not be *exact* equivalents. Readers are advised to consult manufacturers' data before ordering

Recommended cables

Application	Type	Recommended cable
Cheapernet	Coax	Brand Rex GT551014, ICL80049496
Data communications in noisy environments	Twin-axial	Alpha 9817, Belden 9207, IBM7362211
Data communications, low cross-talk	Multipair	Belden 8723, 8777, 8774, etc.
Ethernet trunk	Coax	Belden 9880, BICC8112, NEK06214
Ethernet drop	4-pair, plus drain	Belden 9892, Brand Rex BN-57892, NEK06668
10 base-T	CAT-5	Belden Datatwist 100 CAT-5
100 base-T	CAT-5	Belden Datatwist 350 CAT-5
1000 base-T	CAT-6	Belden Datatwist 350 CAT-6
General-purpose	1-pair (unscreened)	Alpha 1202, Belden 8795, BICC H8150
General-purpose	Multi-pair	Belden 9502, 9504, 9506, etc.
General-purpose	RG62A/U coax	Alpha 9062A, Belden 9269, BICC T3430
General-purpose data/control	Multicore plus coax	Belden 9881
HF radio	URM43 coax	Uniradio M43
Oslan drop	4-pair, plus drain	BICC H960, Brand Rex GT553011, ICL80048808
Point-of-sale terminals	2-pair	Alpha 9819, Belden 9855, BICC H8063, IBM1657265
VHF/UHF radio	URM67 coax	Uniradio M67 (equivalent to RG213/U)

Important note:
Readers are advised to consult manufacturer's data in order to check the suitability of cables before ordering

Optical fibre technology

Optical fibres are becoming widely used as a transmission medium for long-haul data communications and in local area networks (LANs). It is now possible to obtain data rates in excess of 4 Gbps over distances of greater than 100 km and 140 Mbps at distances over 220 km. Submarine cables use optical fibre technology to transmit 160 Gbps over >1000 km. This is achieved using multiple wavelengths and erbium-doped fibre amplifiers.

Optical fibres offer some very significant advantages over conventional waveguides and coaxial cables. These can be summarised as follows:

- Optical cables are lightweight and of small physical size
- Exceptional bandwidths are available within the medium
- Relative freedom from electromagnetic interference
- Significantly reduced noise and cross-talk compared with conventional data cables
- Relatively low values of attenuation within the medium
- High reliability coupled with long operational life
- Electrical isolation and freedom from earth/ground loops
- Very high security of transmission

Optical fibres and their associated high-speed optical sources and detectors are particularly well suited to the transmission of wideband digitally encoded information. This permits the medium to be used for high-speed data communications, local and wide area networking applications.

Propagation

Essentially, an optical fibre consists of a cylindrical glass core surrounded by glass cladding. The fibre acts as a dielectric waveguide in which the electromagnetic wave (of appropriate frequency) will propagate with minimal loss.

Refraction towards the normal

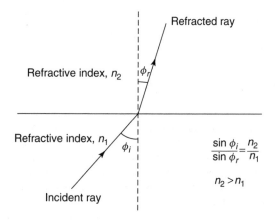

Refraction away from the normal

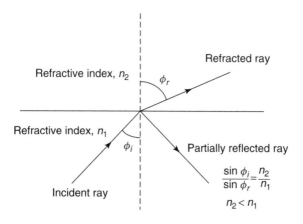

Much of fibre optics is governed by the fundamental laws of refraction. When a light wave passes from a medium of higher refractive index to one of lower refractive index, the wave is bent towards the normal. Conversely, when travelling from a medium of lower refractive index to one of higher refractive index, the wave will be bent away from the normal. In this latter case, some of the incident light will be reflected at the boundary of the two media and, as the angle of incidence is increased, the angle of refraction will also be increased until, at a critical value, the light wave will be totally reflected (ie, the refracted ray will no longer exist). The angle of incidence at which this occurs is known as the critical angle, ϕ_c. The value of ϕ_c depends upon the absolute refractive indices of the media and is given by:

$$\phi_c = \sqrt{\frac{2(n_1 - n_2)}{n_1}}$$

where n_1 and n_2 are the refractive indices of the more dense and less dense media respectively.

Optical fibres are drawn from the molten state and are thus of cylindrical construction. The more dense medium is surrounded by the less dense cladding. Provided the angle of incidence of the input wave is larger than the critical angle, the light wave will propagate along the fibre by means of a series of total reflections. Any other light waves that are incident on the upper boundary at an angle $\phi > \phi_c$ will also propagate along the inner medium. Conversely, any light wave that is incident upon the upper boundary with $\phi < \phi_c$ will pass into the outer medium and will be lost there by scattering and/or absorption.

Launching

Having briefly considered propagation within the fibre, we shall turn our attention to the mechanism by which waves are launched. The cone of acceptance is defined as the complete set of angles which will be subject to total internal reflection. Rays entering from the edges will take a longer path through the fibre but will travel faster because of the lower refractive index of the outer layer. The numerical aperture determines the bandwidth of the fibre and is given by:

$$NA = \sin \phi_a$$

Clearly, when a number of light waves enter the system with differing angles of incidence, a number of waves (or modes) are able to propagate. This multimode propagation is relatively simple to achieve but has the attendant disadvantage that, since the light waves will take different times to pass through the fibre, the variation of transit time will result in dispersion, which imposes an obvious restriction on the maximum bit rate that the system will support.

The cone of acceptance

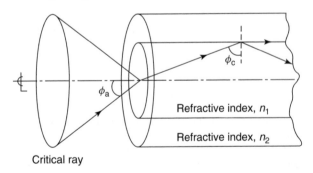

Critical ray

Total internal reflection

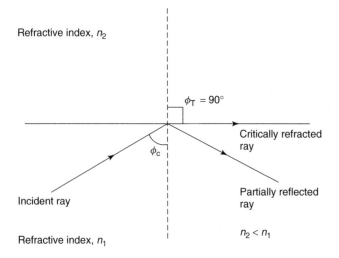

Multimode propagation

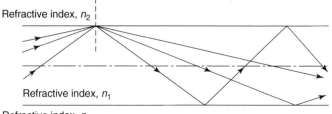

There are two methods for minimising multimode propagation. One uses a fibre of graded refractive index, while the other uses a special monomode fibre. The inner core of this type of fibre is reduced in diameter so that it is of the same order of magnitude as the wavelength of the incident wave. This ensures that only one mode will successfully propagate.

Attenuation

The loss within an optical fibre arises from a number of causes including: absorption, scattering in the core (due to non-homogeneity of the refractive index), scattering at the core/cladding boundary, and losses due to radiation at bends in the fibre.

The attenuation coefficient of an optical fibre refers only to losses in the fibre itself and neglects coupling and bending losses. In general, the attenuation of a good quality fibre can be expected to be approximately 0.3 dB km^{-1} at a wavelength of 1300 nm. Hence a 5 km length of fibre can be expected to exhibit a loss of around 1.5 dB (excluding losses due to coupling and bending). The loss is lower at a wavelength of 1550 nm (typically 0.2 dB km^{-1}) but suffers more from bend induced losses.

Whereas the attenuation coefficient of an optical fibre is largely dependent upon the quality and consistency of the glass used for the core and cladding, the attenuation of all optical fibres varies widely with wavelength. The typical attenuation/wavelength characteristic for a monomode fibre is shown in the figure below. It should be noted that the sharp peak at about 1.39 μm arises from excess absorption within the monomode fibre.

Typical attenuation/wavelength characteristic for a monomode optical fibre

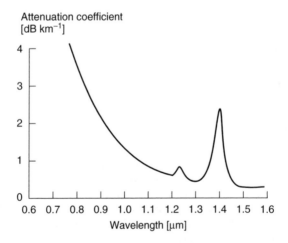

Monomode fibres are now a common feature of high-speed data communication systems and manufacturing techniques have been developed which ensure consistent and reliable products with low attenuation and wide operational bandwidths. However, since monomode fibres are significantly smaller in diameter than their multimode predecessors, a consistent and reliable means of cutting, surface preparation, alignment and interconnection is essential.

Relative dimensions of multimode and monomode fibres

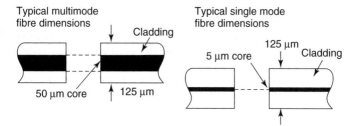

Optical fibre connectors

In long-haul networks the majority of fibre joins are made by fusion splicing two fibres together. This technique uses a small electric arc to create a high temperature and melt the fibre ends. The two ends are pushed together and these bond as the glass cools. Fusion splicing results in a low-loss join, typically <0.1 dB.

At certain points in the fibre network, a 'breakable' join is required and for these a connector must be used. The essential requirements for optical fibre connectors are:

- Low cost
- Robustness
- Repeatability (over numerous mating operations)
- Reliability
- Suitability for installation 'in the field'
- Low loss

There are several connector types in common use. These include SMA, SC, ST and FC. In telecommunications networks FC connectors are common. In LAN networks ST and duplex SC connectors are more common.

Whilst the loss exhibited by a connector may be quoted in absolute terms, it is often specified in terms of an equivalent length of optical fibre. This technique is particularly relevant in the appraisal of long-haul networks. If, for example, two connectors are used at a

repeater, the overall connector loss may approach 1 dB. This is equivalent to several kilometers of low-loss fibre! If the connector loss can
be reduced, then the spacing between repeaters can be increased and
the overall number of repeaters can be reduced accordingly.

While optical fibres are ideal for use in long-haul and wideband
networking applications, they are also suitable for low-speed local
applications where high security and/or reliability of data transfer is
required or where a very high noise level would preclude the use
of conventional cables. A fibre optic RS-232 interface is available
from several manufacturers. This device is fitted with standard SMA
connectors for use with $50/125\,\mu m$ or $200\,\mu m$ optical cables which
operate at a wavelength of 820 nm. SMA terminated optical cables
having lengths of between 2 m and 500 m are available from several
suppliers.

For very short distance applications, inexpensive polymer fibres
may be used. These fibres are generally designed for use at wavelengths of around 665 nm (visible red light); however, since they
generally exhibit attenuation of around $200\,dB\,km^{-1}$, they are only
suitable for short distances (ie, typically less than 50 m).

Optical sources

Suitably mounted and encapsulated light emitting diodes (LED) and
laser diodes (LD) are commonly used as sources in conjunction with
optical fibres. The following table summarises the typical characteristics of these optical sources:

Device type	Material	Operating wavelength (nm)	Bit rate (Mbps)	Transmission range (km)
LED	AlGaAs	850	0 to 40	0 to 5
LED	InGaAsP	1300	0 to 300	5 to 10
LD	InGaAsp	1300	30 to 800	10 to 50
LD	InGaAsp	1550	100 to 800	50 to 100

Higher data rates are usually achieved using an unmodulated laser
and an external electro-absorption modulator (EAM).

Optical detectors

Appropriately mounted and encapsulated photodiodes (PD) or
avalanche photodiodes (APD) are commonly used as detectors in

conjunction with optical fibres. The following table summarises the typical characteristics of these optical detectors:

Device type	Material	Operating wavelength (nm)	Bit rate (Mbps)	Transmission range (km)
PD	Si	850	0 to 30	0 to 2
APD	Si	850	0 to 40	2 to 5
APD	InGaAs	1300	0 to 300	5 to 50
APD	InGaAs	1550	100 to 2500	50 to 100

Optical fibre safety

The human eye is susceptible to damage from laser light and therefore care must be taken when handling the optical fibre used by a working system. NEVER look at the end of a fibre connector of a working system using a microscope or magnifying glass. Indirect viewing should be used, using a camera fitted to a microscope

Optical safety levels were revised during 2000. Most light sources are Safety Class 1 or Class 1M (formerly Class 3A). Class 1 has a 15 mW power limit at 1300 nm and a 10 mW limit at 1550 nm. All LED devices and some lasers come into this category; they are low-power and inherently safe (Class 1). Class 1M covers lasers up to 50 mW power limit at 1300 nm and approximately 150 mW at 1550 nm. These power levels are regarded as safe for 'live working' but only when precautions are taken, as described above.

Handling coated optical fibre is normally quite safe, but handling bare fibre is hazardous because glass fibre becomes brittle when exposed to air. It is quite easy for fibre to pass through the skin, resulting in uncomfortable glass splinters that are very difficult to remove. The handling of bare fibre is normally only necessary when carrying out fusion splicing and this should be done carefully for both safety reasons and for splice quality reasons.

4 Serial interfaces

Serial interfaces are used to connect signals from the personal computer to data communication equipment or other computers using copper cables.

The most common serial interface is RS-232, otherwise known as V.24. This has been used for many years and has been continually enhanced to make sure that it can still provide a valuable function. Most personal computers have at least one RS-232 port. The ITU issued the latest V.24 specification at the end of year 2000.

A number of interface specifications have been used where RS-232 has not been suitable. The most recent are the Universal Serial Bus (USB) and IEEE-1394 (Firewire), which permit very high data rates over copper cables. These interfaces are described here.

Serial data transmission

In serial data transmission one data bit is transmitted after another. In order to transmit a byte of data it is therefore necessary to convert incoming parallel data from the bus into a serial bit stream which can be transmitted along a line.

Serial data transmission can be synchronous (clocked) or asynchronous (non-clocked). The latter method has obvious advantages and is by far the most popular method. The rate at which data is transmitted is given by the number of bits transmitted per unit time. The commonly adopted unit is the 'baud', with 1 baud roughly equivalent to 1 bit per second.

It should, however, be noted that there is a subtle difference between the bit rate as perceived by the computer and the baud rate presented in the transmission medium. The reason is simply that some overhead in terms of additional synchronising bits is required in order to recover asynchronously transmitted data.

In the case of a typical RS-232C link, a total of 11 bits is required to transmit only seven bits of data. A line baud rate of 600 baud thus represents a useful data transfer rate of only some 382 bits per second.

Many modern serial data transmission systems can trace their origins to the 20 mA current loop interface which was once commonly used to connect a teletype unit to the minicomputer system. This system was based on the following logic levels:

$$\text{Mark} = \text{logic } 1 = 20 \text{ mA current flowing}$$

$$\text{Space} = \text{logic } 0 = \text{no current flowing}$$

where the terms 'mark' and 'space' simply refer to the presence or absence of a current.

This system was extended to cater for more modern and more complex peripherals for which voltage, rather than current, levels were appropriate.

Serial I/O devices

Since the data present on a microprocessor bus exists primarily in parallel form (it is *byte wide*) serial I/O is somewhat more complex than parallel I/O. Serial input requires a means of conversion of the parallel data present on the bus into serial output data. In the first case, conversion can be performed with a serial input parallel output (SIPO) shift register whilst in the second case a parallel input serial output (PISO) shift register is required.

Serial data may be transferred in either *synchronous* or *asynchronous* mode. In the former case, all transfers are carried out in accordance with a common clock signal (the clock must be available at both ends of the transmission path). Asynchronous operation involves transmission of data in *packets*; each packet containing the necessary information required to decode the data which it contains. Clearly this technique is more complex but it has the considerable advantage that a separate clock signal is not required. As with programmable parallel I/O devices, a variety of different names are used to describe programmable serial I/O devices but the *asynchronous communications interface adaptor* (ACIA) and *universal asynchronous receiver/transmitter* (UART) are both commonly encountered in data communications.

Signal connections commonly used with serial I/O devices include:

Signal	Function
D0 to D7	Data input/output lines connected directly to the microprocessor bus
RXD	Received data (incoming serial data)
TXD	Transmitted data (outgoing serial data)
CTS	Clear to send. This (invariably active low) signal is taken low by the peripheral when it is ready to accept data from the microprocessor system
RTS	Request to send. This (invariably active low) signal is taken low by the microprocessor system when it is about to send data to the peripheral

As with parallel I/O, signals from serial I/O devices are invariably TTL-compatible. It should be noted that, in general, such signals are unsuitable for anything other than the shortest of transmission paths (eg, between a keyboard and a computer system enclosure). Serial data transmission over any appreciable distance invariably requires additional *line drivers* to provide buffering and level shifting between the serial I/O device and the physical medium. Additionally, *line receivers* are required to condition and modify the incoming signal to TTL levels.

Parallel to serial data conversion

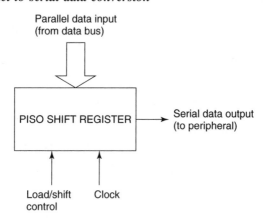

Serial to parallel data conversion

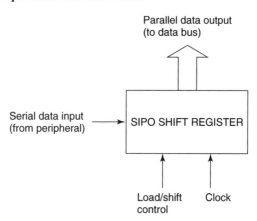

Internal architecture of a representative serial I/O device

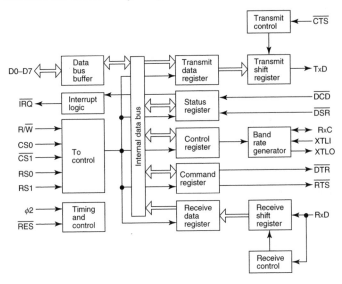

CPU interface to a programmable serial I/O device

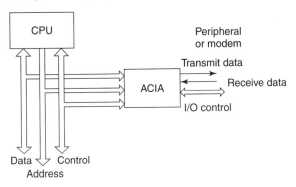

Electrical characteristics of popular interface specifications

Interface specification	Mode	Line	Voltage levels		Logic levels		Data rate (maximum) bps
			min	max	0	1	
V24/V28	asynchronous/ synchronous	unbalanced	3 V	25 V	+ve	−ve	19.2 K
RS-232C	asynchronous/ synchronous	unbalanced	3 V	25 V	+ve	−ve	19.2 K
X26 (V10)	asynchronous/ synchronous	unbalanced	3 V	10 V	+ve	−ve	100 K
RS-423A	asynchronous/ synchronous	unbalanced	0.2 V	6 V	+ve	−ve	100 K
X27 (V11)	asynchronous/ synchronous	balanced	0.3 V	10 V	+ve	−ve	10 M
RS-422A	asynchronous/ synchronous	balanced	0.2 V	6 V	+ve	−ve	10 M
RS485	asynchronous/ synchronous	balanced	0.2 V	6 V	+ve	−ve	10 M
USB2.0	asynchronous/ synchronous	balanced	0.1 V	3.3 V	+ve	−ve	480 M
IEEE1394	synchronous	balanced	118 mV	265 mV	−ve	+ve	400 M

RS-232

The RS-232/ITU-T V.24 interface is widely used for serial communication between microcomputers, peripheral devices, and remote host computers. The RS-232D EIA standard (January 1987) is a revision of the earlier RS-232C standard which brings it in-line with international standards ITU-T V.24, V.28 and ISO IS2110. The RS-232D standard includes facilities for loop-back testing which were not defined under RS-232C.

RS-232 was first defined by the Electronic Industries Association (EIA) in 1962 as a recommended standard (RS) for modem interfacing. The standard relates essentially to two types of equipment; *data terminal equipment* (DTE) and *data circuit-terminating equipment* (DCE).

Data terminal equipment (eg a personal computer) is capable of sending and/or receiving data via an RS-232 serial interface. It is thus said to terminate a serial link. Data circuit terminating equipment (formerly known as *data communications equipment*), on the other hand, is generally thought of as a device which can facilitate serial data communications and a typical example is that of a modem (modulator-demodulator) which forms an essential link in the serial path between a microcomputer and a conventional analogue telephone line.

An RS-232 serial port is usually implemented using a standard 25-way D-connector. Data terminal equipment is normally fitted with a male connector while data circuit-terminating equipment conventionally uses a female connector (note that there are some exceptions to this rule!).

RS-232 signals

RS-232 signals fall into one of the following three categories:

(a) data (eg, TXD, RXD)
RS-232 provides for two independent serial data channels (described as *primary* and *secondary*). Both of these channels provide for full duplex operations (ie, simultaneous transmission and reception).

(b) handshake control (eg, RTS, CTS)
Handshake signals provide the means by which the flow of serial data is controlled allowing, for example, a DTE to open a dialogue with the DCE prior to actually transmitting data over the serial data path.

(c) timing (eg, TC, RC)
For synchronous (rather than the more usual asynchronous) mode of operation, it is necessary to pass clock signals between the devices. These timing signals provide a means of synchronising the received signal to allow successful decoding.

The complete set of RS-232D signals is summarised in the following table, together with EIA and ITU-T designations and commonly used signal line abbreviations.

RS-232D signals and functions

Pin numbers relate to 25-way D-type connectors

Pin number	EIA interchange circuit	ITU-T equiv.	Common abbreviations	Direction	Signal/function
1	–	–	FG	–	frame or protective ground
2	BA	103	TD or TXD	To DCE	transmitted data
3	BB	104	RD or RXD	To DTE	received data
4	CA	105	RTS	To DCE	request to send
5	CB	106	CTS	To DTE	clear to send
6	CC	107	DSR	To DTE	DCE ready
7	AB	102	SG	–	signal ground/common return
8	CF	109	DCD	To DTE	received line signal detector
9	–		–	–	reserved for testing (positive test voltage)
10	–		–	–	reserved for testing (negative test voltage)
11	–		[QM]	–	[Equaliser mode]
12	SCF/CI	122/112	SDCD	To DTE	secondary received line signal detector/data rate select (DCE source)
13	SCB	121	SCTS	To DTE	secondary clear to send
14	SBA	118	STD	To DCE	secondary transmitted data
15	DB	114	TC	To DTE	transmit signal element timing (DCE source)
16	SBB	119	SRD	To DTE	secondary received data
17	DD	115	RC	To DTE	receiver signal element timing (DCE source)
18	LL	141	[DCR]	To DCE	local loop-back [Divided receive clock]
19	SCA	120	SRTS	To DCE	secondary request to send
20	CD	108.2	DTR	To DCE	data terminal ready
21	RL/CG	140/110	SQ	To DCE/ To DTE	remote loop-back/signal quality detector
22	CE	125	RI	To DTE	ring indicator
23	CH/CI	111/112		To DCE/ To DTE	data signal rate selector (DTE)/data signal rate selector (DCE)

Pin number	EIA interchange circuit	ITU-T equiv.	Common abbreviations	Direction	Signal/function
24	DA	113	TC	To DCE	transmit signal element timing (DTE source)
25	TM	142	–	To DTE	test mode

Notes:
1. The functions given in brackets for pin-11 and pin-18 relate to the Bell 113B and 208A specifications
2. Pin-9 and pin-10 are normally reserved for testing. A typical use for these pin numbers is testing of the positive and negative voltage levels used to represent the MARK and SPACE levels
3. For new designs using EIA interchange circuit SCF, CH and CI are assigned to pin-23. If SCF is not used, CI is assigned to pin-12
4. Some manufacturers use spare RS-232 lines for testing and/or special functions peculiar to particular hardware (some equipment even feeds power and analogue signals along unused RS-2320 lines!)

In practice, few RS-232 implementations make use of the secondary channel and, since asynchronous (non-clocked) operation is the norm, only eight or nine of the 25 are regularly used.

Subset of the most commonly used RS-232 signals

Pin numbers relate to 25-way D-type connector

Pin number	EIA interchange circuit	Signal	Function
1	–	FG	earth connection to the equipment frame or chassis
2	BA	TXD	serial data transmitted from DTE to DCE
3	BB	RXD	serial data received by the DTE from the DCE
4	CA	RTS	when active, the DTE is signalling that it wishes to send data to the DCE
5	CB	CTS	when active, the DCE is signalling that it is ready to accept data from the DTE
6	CC	DSR	when active, the DCE is signalling that a communications path has been properly established
7	AB	SG	common signal return path
8	CF	DTR	when active, the DTE is signalling that it is operational and that the DCE may be connected to the communications channel

The table above gives the subset of commonly used RS-232 signals. Note that the pin numbers relate to the 25-way connector. In the 9-way connector, this subset is used but the pin numbers are different.

The most confusing difference between 25-way and 9-way connectors is that the functionality of pin numbers 2 and 3 are interchanged. Please compare the later diagrams 'V.24/RS-232 interface connections' for the 25-way variety with the 'V.24/RS-232 (9-pin) interface connections' for the 9-way variety.

RS-232 waveforms

In most RS-232 systems, data is transmitted asynchronously as a series of packets, each representing a single ASCII character and containing sufficient information for it to be decoded without the need for a separate clock signal.

ASCII characters are represented by seven binary digits (bits). The upper case letter *A*, for example, is represented by the seven-bit binary word; 1000001. In order to send the letter *A* via an RS-232 system, we need to add extra bits to signal the start and end of the data packet. These are known as the *start bit* and *stop bit* respectively. In addition, we may wish to include a further bit to provide a simple parity error detecting facility.

One of the most commonly used schemes involves the use of one start bit, one parity bit, and two stop bits. The commencement of the data packet is signalled by the start bit which is always low irrespective of the contents of the packet. The start bit is followed by the seven data bits representing the ASCII character concerned. A parity bit is added to make the resulting number of 1s in the group either odd (*odd parity*) or even (*even parity*). Finally, two stop bits are added. These are both high.

The complete asynchronously transmitted data word thus comprises eleven bits (not that only seven of these actually contain data). In binary terms the word can be represented as: 01000001011. In this case, even parity has been used and thus the ninth (*parity bit*) is a 0.

Voltage levels employed in an RS-232 interface are markedly different from those used within a microcomputer system. A positive voltage (of between +3 V and +25 V) is used to represent a logic 0 (or *space*) while a negative voltage (of between −3 V and −25 V) is used to represent a logic 1 (or *mark*).

Level shifting (from TTL to RS-232 signal levels and vice versa) is invariably accomplished using *line driver* and *line receiver* chips, the most common examples being the 1488 and 1489 devices.

*Typical representation of the ASCII character **A** using TTL signal levels*

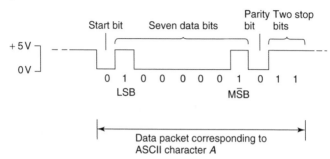

*ASCII character **A** as it appears on TD or RD signal lines*

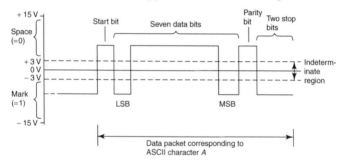

RS-232 electrical characteristics

The following summarises the principal electrical specification for the RS-232 standard:

Maximum line driver output voltage (open circuit):	± 25 V
Maximum line driver output current (short circuit):	± 500 mA
Minimum line impedance: $3\,k\Omega$ in parallel with $2.5\,nF$	
Line driver space output voltage ($3\,k\Omega \le R_L \le 7\,k\Omega$):	+5 V to +15 V
Line driver mark output voltage ($3\,k\Omega \le R_L \le 7\,k\Omega$):	−5 V to −15 V
Line driver output (idle state):	mark
Line receiver output with open circuit input:	logic 1
Line receiver output with input ≥ 3 V:	logic 0
Line receiver output with input ≥ -3 V:	logic 1

Maximum transition times are defined as follows:

Unit interval (UI)	Maximum transition time
≥25 ms	1 ms
25 ms to 125 μs	4% of UI
less than 125 μs	5 μs

RS-232 logic and voltage levels

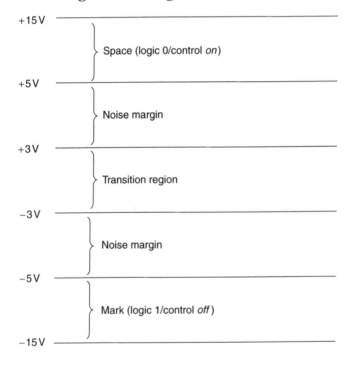

Simplified arrangement of a microcomputer RS-232 interface

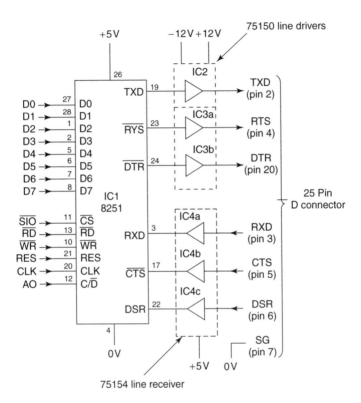

IC1 is a programmable serial I/O device while IC2 and IC3 provide level shifting and buffering for the three output signals (TXD, RTS and DTR). IC4 provides level shifting for the three input signals (RXD, CTS and DSR). Note that IC2 and IC3 both require ±12 V supplies and that mark and space will be represented by approximate voltage levels of −12 V and +12 V respectively.

RS-232 data cables

(a) 4-way cable for dumb terminals

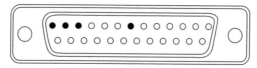

Pins used: 1–3 and 7 (pins 8 and 20 are jumpered)

(b) 9-way cable for asynchronous communications

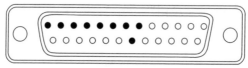

Pins used: 1–8 and 20

(c) 15-way cable for synchronous communications

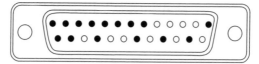

Pins used: 1–8, 13, 15, 17, 20, 22 and 24

(d) 25-way cable for universal applications

Pins used: 1–25

Male and female 25-way D-connectors used for RS-232

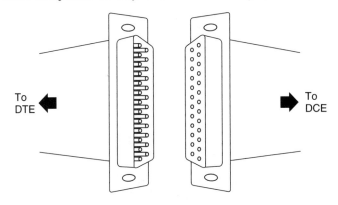

RS-232 pin connections

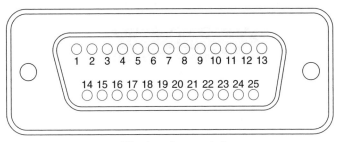

(Pin view of connector)

V.24/RS-232 interface connections

Source	Signal designation		Signal designation	Source
DTE	Secondary transmit data	14	1 Protective ground	Common
DCE	Transmit clock (DCE source)	15	2 Transmit data, TXD	DTE
DCE	Secondary received data	16	3 Receive data, RXD	DCE
DCE	Receive clock	17	4 Request to send, RTS	DTE
DTE	Local loopback, LL	18	5 Clear to send, CTS	DCE
DTE	Secondary request to send	19	6 Data set ready, DSR	DCE
DTE	Data terminal ready, DTR	20	7 Signal ground	Common
DTE	Remote loopback, RL	21	8 Carrier detect	DCE
DCE	Ring indicator, RI	22	9 Reserved (+V)	n/a
DTE/DCE	Baud rate select	23	10 Reserved (−V)	n/a
DTE	Transmit clock (DTE source)	24	11 Unassigned	n/a
DCE	Test mode	25	12 Secondary carrier detect	DCE
			13 Secondary clear to send	DCE

V.24/RS-232 (9-pin) interface connections

Source	Signal designation		Signal designation	Source
			1 Data carrier detect, DCD	DCE
			2 Receive data, RXD	DCE
			3 Transmit data, TXD	DTE
			4 Data terminal ready, DTR	DTE
			5 Ground, GND	Common
DCE	Data set ready, DSR 6			
DTE	Request to send, RTS 5			
DCE	Clear to send, CTS 4			
DCE	Ring indicator, RI 3			

RS-232 enhancements

Several further standards have been introduced in order to overcome some of the shortcomings of the original RS-232 specification. These generally provide for better line matching, increased distance capability, and faster data rates. Notable among these systems are RS-422 (a balanced system which caters for a line impedance as low as 50 ohm), RS-423 (an unbalanced system which will tolerate a line impedance of 450 ohm minimum), and RS-449 (a very fast serial data standard which uses a number of changed circuit functions and a 37-way D connector).

RS-422

RS-422 is a balanced system (differential signal lines are used) which employs lower line voltage levels than those employed with RS-232. Space is represented by a line voltage level in the range $+2$ V to $+6$ V while mark is represented by a line voltage level in the range -2 V to -6 V. RS-422 caters for a line impedance of as low as 50 ohm and supports data rates of up to 10 Mbps.

RS-422 logic and voltage levels

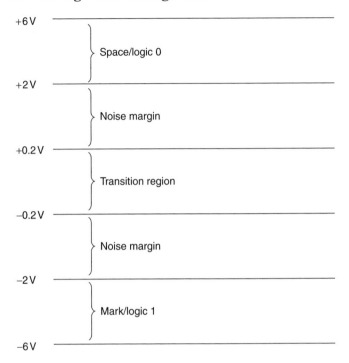

RS-423

Unlike RS-422, RS-423 employs an unbalanced line configuration (a single signal line is used in conjunction with signal ground). Line voltage levels of +4 V to +6 V and −4 V to −6 V represent space and mark respectively and the standard specifies a minimum line terminating resistance of 450 ohm. RS-423 supports a maximum data rate of 100 kbps.

RS-423 logic and voltage levels

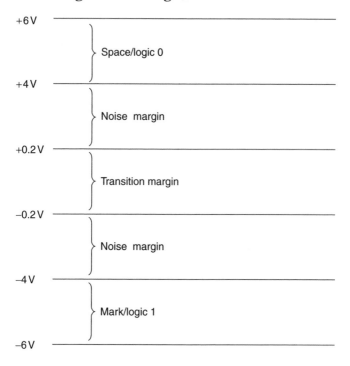

RS-449

RS-449 is a further enhancement of RS-422 and RS-423 which caters for very fast data rates (up to 2 Mbps) yet provides for upward compatibility with RS-232. Ten extra circuit functions have been provided while three of the original interchange circuits have been abandoned. In order to minimise confusion, and since certain changes have been made to the definition of circuit functions, a completely new set of circuit abbreviations has been developed. In addition, the standard requires 37-way and 9-way D-connectors, the latter being necessary where use is made of the secondary channel interchange circuits.

RS-449 pin connection and interchange circuits

Auxiliary connector (9-way)	Main connector (37-way)		Circuit abbreviation	Function
	A	B		
1	1			shield
5	19		SG	signal ground
9	37		SC	send common
6	20		RC	receive common
	4	22	SD	send data
	6	24	RD	receive data
	7	25	RS	request to send
	9	27	CS	clear to send
	11	29	DM	data mode
	12	30	TR	terminal ready
	15		IC	incoming call
	13	31	RR	receive ready
	33		SQ	signal quality
	16		SR	signalling rate selector
	2		SI	signalling rate indicator
	17	35	TT	terminal timing
	5	23	ST	send timing
	8	26	RT	receive timing
3			SSD	secondary send data
4			SRD	secondary receive data
7			SRS	secondary request to send
8			SCS	secondary clear to send
2			SRR	secondary receiver ready
	10		LL	local loop-back
	14		RL	remote loop-back
	18		TM	test mode
	32		SS	select standby
	36		SB	standby indicator
	18		SF	select frequency
	28		IS	terminal in service
	34		NS	new signal

Notes:

1. Pins 3 and 31 of the 37-way connector are undefined
2. *B* on the main connector indicates a return signal

RS-449 pin connections

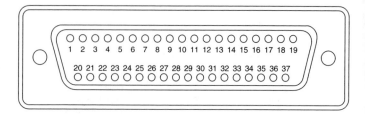

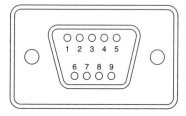

(Pin view of connectors)

V.36/RS-449 interface connections

Source	Signal designation		Signal designation	Source
Common	Send common 37		19 Signal ground	Common
DCE	Standby indicator 36		18 Test mode	DCE
Return	Terminal timing (B) 35		17 Terminal timing (A)	DTE
DTE	New signal 34		16 Select frequency	DTE
DTC	Signal quality 33		15 Incoming call	DCE
DTE	Select standby 32		14 Remote loopback	DTE
Return	Receiver ready (B) 31		13 Receiver ready (A)	DCE
Return	Terminal ready (B) 30		12 Terminal ready (A)	DTE
Return	Data mode (B) 29		11 Data mode (A)	DCE
DTE	Terminal in service 28		10 Local loopback	DTE
Return	Clear to send (B) 27		9 Clear to send (A)	DCE
Return	Receive timing (B) 26		8 Receive timing (A)	DCE
Return	Request to send (B) 25		7 Request to send (A)	DTE
Return	Receive data (B) 24		6 Receive data (A)	DCE
Return	Send timing (B) 23		5 Send timing (A)	DCE
Return	Send data (B) 22		4 Send data (A)	DTE
n/a	Unassigned 21		3 Unassigned	n/a
Common	Receive common 20		2 Signal rate indicator	DCE
			1 Shield	Common

RS-485

The Electronic Industries Association (EIA) and the Telecommunications Industry Association (TIA) jointly developed the RS-485 standard. Strictly, this should now be referred to as EIA/TIA-485.

RS-485 is closely related to the RS-422 standard, with balanced line transmission. RS-422 has one driver and a number of receivers: the driver is always active on the line. But RS-485 allows bi-directional half-duplex operation, with one driver connected at each end of the cable pair. Driver and receiver pairs can be located at various points along the bus, but only one driver can be active at any one time. The unused driver is put into the high-impedance state, so that it does not affect the data signals transmitted from the other end of the link. Parts intended for RS-485 can also be used for RS-422, but the reverse is not true because RS-422 drivers cannot relinquish control of the bus.

The bus should be one continuous pair with a 120-ohm terminating resistor at either end. An alternative 'fail-safe' bus termination has a 130-ohm terminating resistor, a 750-ohm 'pull-up' resistor from the A-wire to +5 V and a 750-ohm 'pull-down' resistor from the B-wire to ground. Spurs off the bus should not be allowed, unless the system operates at low-speed, because reflections from an unterminated spur will affect the data pulse shape and cause errors. The bus can be up to 1250 metres long, at data rates of up to 100 kbps. At higher data rates the maximum line length is reduced. At 10 Mbps, the maximum line length is about 30 metres.

RS-485 logic levels and connectors

The RS-485 logic and voltage levels are the same as those previously given for RS-422. Each receiver has a maximum threshold of 200 mV. The minimum output level from any driver is 1.5 V.

There is no connector, cable or protocol specification for RS-485.

RS-530 interface connections

Source	Signal designation	Pin		Pin	Signal designation	Source
DCE	Test mode	25		13	Clear to send (B)	Return
DTE	Ext. transmit clock (A)	24		12	Transmit clock (B)	Return
Return	DTE ready (B)	23		11	Ext. transmit clock (B)	Return
Return	DCE ready (B)	22		10	Receive line signal detector (B)	Return
DTE	Remote loopback	21		9	Receive clock (B)	Return
DTE	DTE ready (A)	20		8	Receive line signal detector (A)	DCE
Return	Request to send (B)	19		7	Signal ground	Common
DTE	Local loopback	18		6	DCE ready	DCE
DCE	Receive clock (A)	17		5	Clear to send (A)	DCE
Return	Receive data (B)	16		4	Request to send (A)	DTE
DCE	Transmit clock (A)	15		3	Receive data (A)	DCE
Return	Transmit data (B)	14		2	Transmit data (A)	DTE
				1	Shield	Common

X.21 interface connections

Source	Signal designation		Signal designation	Source
DTE	Transmit (B) 9		1 Shield	n/a
DTE	Control (B) 10		2 Transmit (A)	DTE
DCE	Receive (B) 11		3 Control (A)	DTE
DCE	Indication (B) 12		4 Receive (A)	DCE
DCE	Signal timing (B) 13		5 Indication (A)	DCE
n/a	Unassigned 14		6 Signal timing (A)	DCE
n/a	Unassigned 15		7 Unassigned	n/a
			8 Ground	Common

V.35 interface connections

Source	Signal designation		Signal designation		Source
Common	Signal ground	B	A	Chassis ground	Common
DCE	Clear to send	D	C	Request to send	DTE
DCE	Data carrier detect	F	E	Data set ready	DCE
DCE	Ring indicator	J	H	Data terminal ready	DTE
n/a	Unassigned	L	K	Unassigned	n/a
n/a	Unassigned	N	M	Unassigned	n/a
DCE	Receive data (A)	R	P	Transmit data (A)	DTE
DCE	Receive data (B)	T	S	Transmit data (B)	DTE
DCE	Receive timing (A)	V	U	Terminal timing (A)	DTE
DCE	Receive timing (B)	X	W	Terminal timing (B)	DTE
n/a	Unassigned	Z	Y	Transmit timing (A)	DCE
n/a	Unassigned	BB	AA	Transmit timing (B)	DCE
n/a	Unassigned	DD	CC	Unassigned	n/a
n/a	Unassigned	FF	EE	Unassigned	n/a
n/a	Unassigned	JJ	HH	Unassigned	n/a
n/a	Unassigned	LL	KK	Unassigned	n/a
n/a	Unassigned	NN	MM	Unassigned	n/a

RJ-11 interface connections

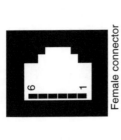

Female connector

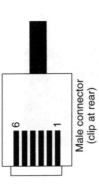

Male connector
(clip at rear)

Pin	Signal
6	n/c
5	Receive +
4	Transmit +
3	Transmit −
2	Receive −
1	n/c

NB: 1 and 6 may be grounded
in some cases

RJ-12 interface connections

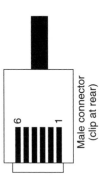

Female connector

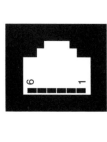

Male connector
(clip at rear)

Pin	Signal
6	Shield
5	Receive +
4	Transmit +
3	Transmit −
2	Receive −
1	Shield

RJ-45 interface connections

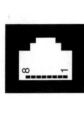

Female connector

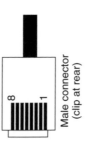

Male connector
(clip at rear)

Pin Signal

8 Clear to send, CTS
7 Signal ground, GND
6 Data set ready, DSR
5 Transit data, TXD
4 Data carrier detect, DCD
3 Receive data, RXD
2 Request to send, RTS
1 Chassis ground

NB: The above relates to the most common
X.25/RS-232 implementation of
this connector

USB

The purpose of USB is to replace most of the traditional ports on a PC a single versatile and user-friendly interface. USB is a shared serial bus using defined protocols. Most of the interface intelligence is placed in the host computer, allowing the use of less complex and less expensive peripherals. It was intended to be a desktop bus for standard peripherals, but USB is now an option for most devices that were previously connected by an RS-232 or parallel port.

There are two versions of USB. Version 1.x supports two bus speeds: **full-speed** at 12 Mbps and **low-speed** at 1.5 Mbps. Low-speed data rates target cost-sensitive peripherals, as well as mice and other devices that require unshielded flexible cables. USB 2.0 also supports **high-speed** at 480 Mbps, which opens the interface to new applications.

Physically, the bus comprises four wires: two for power and two for data. The USB connector therefore has four pins. Pin 1 connects to +5 V, pin 2 connects to D+, pin 3 connects to D− and pin 4 connects to 0 V (ground). The PC host and self-powered peripherals (such as a printer) provide a current limited power source. Thus peripherals (such as a PC's mouse) can be powered over the USB bus. By using all four wires for each peripheral, power feeding the network of non-powered peripherals is shared between the powered devices. Also, using a PC as a power source, a non-powered keyboard can relay power to attached peripherals such as a mouse and a joystick.

The cable used to connect low-speed devices in a USB version 1.1-compliant system requires no shielding. If a USB 2.0 interface is low-speed, it must meet new shielding requirements. A USB 2.0-compliant low-speed cable must have the same aluminium-metallised-polyester inner shield and copper drain wire required for full- and high-speed cables. A braided outer shield and a twisted pair for data are also recommended, as used on full- and high-speed cables.

High-speed USB 2.0 buses allow the use of low- and full-speed devices while transferring data at high-speed whenever possible. On a full-speed bus, the host controller divides the bus time into frames, each 1 ms long. Every frame begins with an SOF (start-of-frame) packet that devices use as a timing reference. Within each frame, the host can schedule multiple transactions to multiple destinations. Each transaction includes an endpoint address that identifies the device buffer to be used. In most transaction types, information travels in both directions. The host initiates the transaction, data travels to or from the host, and then the receiving destination returns status information.

For high-speed traffic, the host divides each frame into eight microframes, each beginning with an SOF packet. Each microframe can carry multiple transactions to multiple destinations. Compared with full-speed, individual transactions can carry more data. USB version 2.0 protocol enhancements make better use of the bus at all speeds.

Hubs can increase the number of ports available to peripherals. A typical hub has one upstream port that transmits toward and receives from the host and as many as seven downstream ports that can connect to peripherals or additional hubs. A 1.x hub supports both low- and full-speed data rates but does not convert between speeds; it just passes the traffic on, changing only the edge rate to match the destination's speed. A 2.0 hub acts as a remote processor and converts from high-speed data to low- or full-speed data as needed.

A USB version 1.x hub determines a device's speed by detecting the voltages on the D+ and D− signal wires from the device. If D− is pulled up at the device, its USB interface is low-speed, and the hub passes only low-speed traffic to the device. In the other direction, the hub passes all the traffic that it receives from the low-speed device to the host. If D+ is pulled up at the device, the device's USB interface is full-speed, and the hub passes low- and full-speed traffic in both directions.

If a low- or full-speed device is connected to a hub that is receiving high-speed data from upstream, a transaction translator in the hub converts the data to low- or full-speed before passing it on. In the other direction, the hub converts low- or full-speed data to high speed before sending it toward the host. To reduce jitter, the hub re-synchronises received high-speed data but otherwise passes it unchanged to any attached high-speed devices.

USB supports four data transfer types:

- Control transfers are for enumeration and other times when the host wants to send defined requests and (optionally) receive data in reply.
- Interrupt transfers are for pointing devices and other applications that need to transfer data at intervals, with a guaranteed maximum time between transactions.
- Bulk transfers are for printers, scanners, and other devices that would like to transfer data as quickly as possible but can wait if the bus is busy.
- Isochronous transfers are for real-time audio and other applications that require guaranteed delivery time but need no error correcting in the transfer.

A USB transaction consists of two or more packets. To begin a transaction, the host sends a token packet containing information about the transaction. A transaction must also have a data packet, a handshake packet where the receiver of the data returns status information, or both. (If no data packet exists, the device sends the status information.)

In addition to the high-speed bit rate, USB 2.0 has an improved protocol for high-speed transfers and new split transactions for full- and low-speed transfers on high-speed buses. For bulk and isochronous transfers, high-speed transfers are about 40 times faster than full-speed transfers are, simply because the bus speed is 40 times faster. But a high-speed interrupt transfer can be almost 400 times faster. Two reasons exist for this speed increase: The maximum packet size per transaction is much greater, and a transaction can have multiple data packets in a frame. High-speed control transfers are also much faster because they can transfer more data per frame.

For high-speed bulk and control transfers, 2.0 supports an improved protocol that uses less bus time to determine whether a device is ready to receive data. With full- and low-speed devices, when the host wants to send data in a control, bulk, or interrupt transfer, it sends the token and data packets and receives a reply from the device in the handshake packet of the transaction. If the device isn't ready for the data, it returns a NAK (negative acknowledgment), and the host tries again later. This protocol can waste a lot of bus time if the device is rarely ready.

High-speed bulk and control transactions have a better handshaking method. After receiving data, a device endpoint can return a NYET (not yet) handshake, which says that the endpoint accepted the data, but it is not yet ready to receive more data. When the host thinks the device might be ready, it sends a PING token packet, and the endpoint returns an ACK (acknowledge) or a NAK to indicate whether the host can send the next transaction's data.

To efficiently use bus time, high-speed hosts and hubs use new split transactions with low- and full-speed devices. At low- and full-speed, all of a transaction's packets are in sequence, with no other traffic between them. For example, on receiving token and data packets, a device must return an expected handshake packet without delay. But a high-speed hub could waste a lot of time waiting for a low- or full-speed device to receive the token and data packets and return a response.

Two part transactions are used to reduce wasted time. At high-speeds, the host sends the hub a start-split token packet along with any data the host is sending in the transaction. The host is then free to do other transactions without waiting for this transaction to complete. The high-speed hub then translates to low- or full-speed and completes the transaction with the destination device. However, instead of the hub passing on the device's response to the host immediately upon receipt, it stores the response in a buffer. Later, the host sends a complete-split token packet to request the device's response from the hub. The hub returns the data or handshake packet and completes the transaction with the host.

USB receivers detect a differential 0 or 1 by measuring whether the D+ or D− input is more positive. The voltage on each line must also be within a specified, absolute range. Transceivers must have separate drivers for high speed. For receiving, transceivers may have one pair of receivers that handles all speeds or separate receivers for high speeds.

In a high-speed driver, a current source drives one line, with the other line at ground. To conserve power, a high-speed driver can activate its current source only when transmitting. This approach complicates the design, however, because the spec requires the device to meet amplitude and timing requirements from the very first symbol in a packet. So the spec also allows the driver to keep its current source active at all times, directing the current to ground when the device is not transmitting on the bus.

In a transceiver that is capable of high-speed data rates, the output impedance of the full-speed drivers has less tolerance (45 ohm, ±10% compared with 36 ohm, ±22%). The change is necessary because the high-speed bus uses the full-speed drivers to terminate the line.

When the high-speed drivers are active, the full-speed drivers bring both data lines low (USB's single-ended zero state), resulting in each driver and its series resistor providing a 45 ohm termination to ground. These terminations at both the source and load quiet the line more effectively than the source-only terminations on a full-speed cable segment. The series resistors may be on- or off-chip. Drivers that aren't part of a high-speed transceiver require no changes in output impedance.

In a low- or full-speed device, a 1.5-kohm pull-up resistor on one of the signal lines indicates device speed. Both wires also have 15-kohm pull-down resistors at the hub. At high speeds, the pull-downs remain, but not the pull-up. When a device switches to high speed, it must remove the pull-up from the line.

When you attach a low- or full-speed device to the bus or remove one from the bus, the voltage change due to the pull-up informs the hub of the change. High-speed devices always attach at full speed, so the hub detects these devices in the usual way.

The switch to high speed occurs during the reset signal, which the hub sends after it detects the device. A device that is capable of supporting high-speed data rates must support the new high-speed handshake that informs the hub that the device can handle high speeds and switches to high speed if possible.

The hub must also detect the removal of a high-speed device, which has no pull-up. It does so by checking the voltage during the EOP (end-of-packet) signal in each high-speed SOF packet. When you remove a device from the bus, you remove its differential terminations, doubling the voltage at the hub's port. When the hub detects the doubled voltage, it knows that the device has been removed.

IEEE-1355

IEEE-1355 is an all-purpose inter-connect standard and is intended for short distances (tens of metres) like RS-232. It requires a UART type device and can operate at data rates from 1 Mbps to 1 Gbps, or more.

IEEE-1355 uses ATM-like packets with an address header to describe the path or channel required. The routing of each packet through the network uses a packet switch.

IEEE-1394 'Firewire'

USB's high-speed competitor is IEEE-1394, also known as Firewire. IEEE-1394 has a bus speed of 400 Mbps, and IEEE-1394b proposes to increase this rate to 3.2 Gbps. Note that the two buses have different purposes, although some peripherals could use either device. With USB, the host initiates every transfer, and every transfer has one destination. With IEEE-1394, peripherals can communicate directly with each other, and a transfer can have multiple destinations. IEEE-1394 devices require more intelligence to manage their communications, so their peripheral controllers are more complex and expensive.

The IEEE-1394 bus uses special connectors with six connection pins. The first two pins are for power: Pin 1 is V+ and pin 2 is ground. Pins 3 and 4 are for transmit data (strobe on receive). Pins 5 and 6 are for receive data (strobe on transmit). Thus pins 3 and 4 at one end of a cable are connected to pins 5 and 6, respectively, at the other end of the cable.

The IEEE-1394 bus is limited to connecting 63 devices. Each device has multiple bus connections, allowing devices to communicate via the ports of other devices. The cable sections between any two devices can be up to 4.5 metres long. There is a maximum limit of 16 cable hops between any two devices on the network. An example of an IEEE-1394 bus connection is shown in the following diagram. In this diagram, device C communicates to device B via the ports of devices A and B. Three cable hops are used to make this connection.

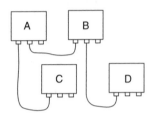

5 Data communication equipment

Data communication equipment (DCE) describes any equipment carrying data communications between terminals. Data terminal equipment (DTE) may be a personal computer, server or workstation. Typically DCE includes modems for transmitting data over telephone lines, and local area network (LAN) equipment such as routers, bridges, hubs and gateways.

Typical links between computers

Typical link between a microcomputer and a local host computer (both configured as DTE)

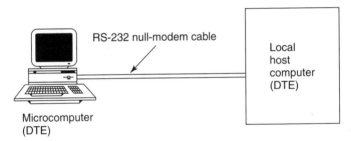

Microcomputer
(DTE)

Typical link between a microcomputer and a remote host computer (both configured as DTE)

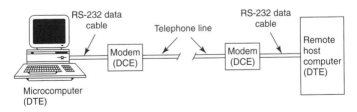

Microcomputer
(DTE)

Typical link between two microcomputers (both configured as DTE)

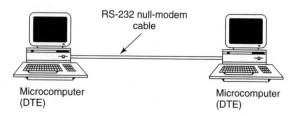

Microcomputer
(DTE)

Microcomputer
(DTE)

Typical null-modem arrangements

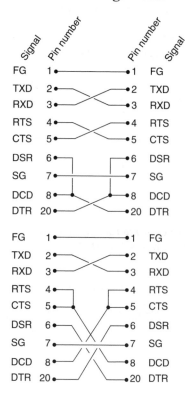

Modems

The name modem is a contraction of modulator–demodulator and this succinctly describes the function of a device that has the dual role of:

- Modulating an outgoing baseband signal onto a carrier for transmission through a physical medium
- Demodulating an incoming modulated carrier from the physical medium in order to recover an input baseband signal.

The modulation method employed in the simplest modems (such as V.21) is frequency shift keying (FSK). A sinusoidal signal of one frequency represents a space (logic 0) and that another frequency represents a mark (logic 1). The frequencies used for the mark and space tones are chosen so that they can be passed through the transmission medium with minimal attenuation. Hence, in the case of modems used with ordinary telephone lines, both mark and space must be represented by audible tones in the frequency range 300 Hz to 3.4 kHz.

The available bandwidth within the transmission medium (telephone line) also has a bearing upon the signalling rate. A wider bandwidth will permit signalling (i.e. switching between mark and space tones) at a faster rate. In practice, the maximum signalling rate for FSK modems transmitting over a conventional exchange line is in the region of 1300 baud. It is, however, possible for a single exchange line to support duplex working in which case different mark and space frequencies must be employed for transmit and receive.

Filters are used within FSK modems to separate transmit and receive frequencies and each end of the link must employ a different pair of mark and space frequencies. The frequencies used for setting up a data transfer (i.e. those used for originate mode) will thus be different from those which are used in response to such a request (i.e. those used in answer mode). When communication is established with a larger remote host, the user-modem will normally establish the call in originate mode.

Medium-speed modems use more complex modulation methods. Phase modulation is used by V.22/Bell 212 (1200 baud full duplex) and V.26/Bell 201 (2400 baud full duplex) modems. In both types of modem an 1800 Hz carrier signal is phase modulated, with 90, 180 or 270 degree phase shifts to indicate the data. Each phase shift signals the state of two data bits.

A slightly higher speed modem standard is V.27, which operates at 4800 baud. A V.27 modem uses one of eight possible phase shifts, each 45 degrees apart, to signal three data bits. Thus no phase change represents 001, a 45 degree change represents 000, a 90 degree change represents 010, etc.

The highest speed telephone line modems use both amplitude and phase modulation (known as quadrature amplitude modulation, or QAM) with trellis encoding. V.34 modems operate at up to 33.6 kbps. A variant on V.34 is V.90, which uses DC levels from the exchange rather than tones to signal the remote modem. Each DC level is used to indicate up to 7 data bits. A V.90 modem uses V.34 signalling in the remote modem to exchange direction.

In medium and high-speed modems, both transmit and receive data are transmitted using the same 1800 Hz carrier frequency. In order to allow full duplex working, echo cancellation is used in the modem to separate received signals from locally generated signals. Thus only the carrier signal received over the telephone line enters the demodulator circuit.

Signal frequencies are governed by a number of internationally agreed (ITU-T) standards in the V-series. It is rare for low-speed modems like V.21 to be used, but V.34 and V.90 are popular. The latest high-speed modem is V.92, with 56 kbps from the exchange and up to 44 kbps from the modem. Most modern modems are able to support a number of standards as well as providing auto-originate/auto-answer facilities. A modem will sometimes revert to a lower speed if the transmission path cannot support the highest speeds.

Communications software is normally required to set up the serial port to which the modem is connected (via RS-232) and, in many cases also to configure the modem. Software will generally provide for a range of signalling speeds (baud rates) and handshaking protocols (e.g. X-ON/X-OFF).

Simplified block schematic of a modem

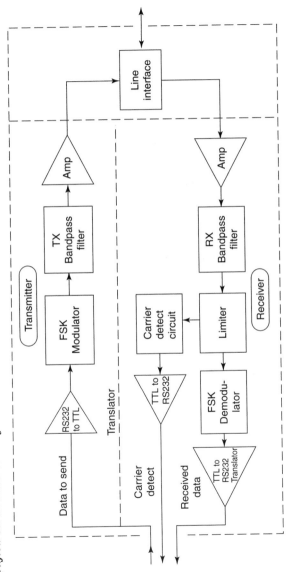

Permitted transmitted spectrum (BS 6305)

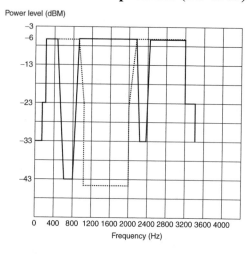

Modem signal frequencies

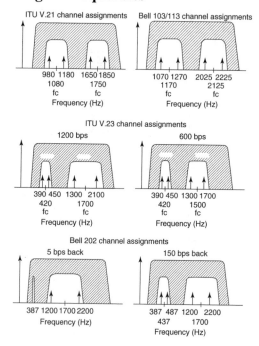

Frequency parameters

Modem	Baud Rate (BPS)	Duplex	Transmit Frequency		Receive Frequency		Answer tone Freq Hz	Soft Turn Off Tone Hz
			Space Hz	Mark Hz	Space Hz	Mark Hz		
Bell 103 Originate	300	Full	1070	1270	2025	2225	–	–
Bell 103 Answer	300	Full	2025	2225	1070	1270	2225	–
CCITT V.21 Originate	300	Full	1180	980	1850	1650	–	–
CCITT V.21 Answer	300	Full	1850	1650	1180	980	2100	–
CCITT V.23 Mode 1	600	Half	1700	1300	1700	1300	2100	900
CCITT V.23 Mode 2	1200	Half	2100	1300	2100	1300	2100	900*
CCITT V.23 Mode 2 Equalised	1200	Half	2100	1300	2100	1300	2100	900*
Bell 202	1200	Half	2200	1200	2200	1200	2025	900
Bell 202 Equalised	1200	Half	2200	1200	2200	1200	2025	900
CCITT V.23 Back	75/150	–	450	390	450	390	–	–
Bell 202 150 bps Back	150	–	487	387	487	387	–	–

Note: *For V.23 soft turn off modes only.

V.21 frequency spectrum (300/300 baud)

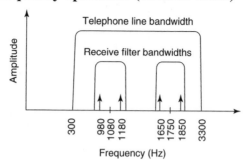

V.21 channels for 300/300 baud

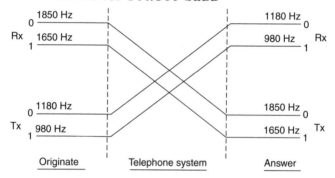

Transmitter carrier and signalling frequency specifications

Frequency	Specification (Hz ±0.01%)
V.22 bis low channel, originate mode	1200
V.22 low channel, originate mode	1200
V.22 high bis channel, answer mode	2400
V.22 high channel, answer mode	2400
Bell 212A high channel, answer mode	2400
Bell 212A low channel, originate mode	1200
Bell 103/113 originating mark	1270
Bell 103/113 originating space	1070
Bell 103/113 answer mark	2225
Bell 103/113 answer space	2025
V.34 originate and answer mode	1800

Line signal encoding (V.26A and V.26B)

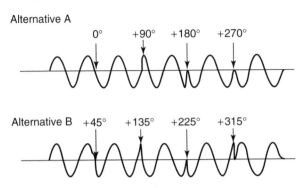

Alternative A

0°   +90°   +180°   +270°

Alternative B   +45°   +135°   +225°   +315°

Differential two-phase encoding (V26bis)

1200 bps	
Bit	Phase change
0	+90°
1	+270°

Differential four-phase encoding (V.26A and V.26B/Bell 201)

2400 bps		
	Phase change	
Dibit	V.26A	V.26B/Bell 201
00	0°	+45°
01	+90°	+135°
11	+180°	+225°
10	+270°	+315°

Typical bit error rate performance for a modem

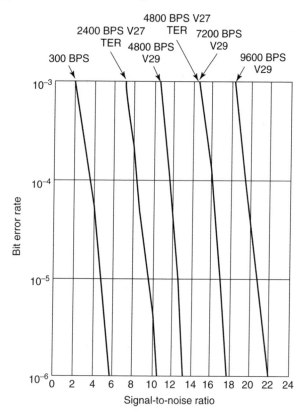

Data communications test equipment

A number of specialised test instruments and accessories are required for testing data communications systems. The following items are commonly encountered.

Patch boxes

These low-cost devices facilitate the cross connection of RS-232 (or equivalent) signal lines. The equipment is usually fitted with two D-type connectors (or ribbon cables fitted with a plug and socket) and all lines are brought out to a patching area into which links may be plugged. In use, these devices are connected in series with the RS-232 serial data path and various patching combinations are tested until a functional interface is established. If desired, a dedicated cable may then be manufactured in order to replace the patch box.

Gender changers

Gender changers normally comprise an extended RS-232 connector which has a male connector at one end and a female connector at the other. Gender changers permit mixing of male and female connector types (note that the convention is male at the DTE and female at the DCE).

Null modems

Like gender changers, these devices are connected in series with an RS-232 serial data path. Their function is simply that of changing the signal lines so that a DTE is effectively configured as a DCE. Null modems can easily be set up using a patch box or manufactured in the form of a dedicated null-modem cable.

Line monitors

Line monitors display the logical state (in terms of mark or space) present on the most commonly used data and handshaking signal lines. Light emitting diodes (LED) provide the user with a rapid indication of which signals are present and active within the system.

Breakout boxes

Breakout boxes provide access to the signal lines and invariably combine the features of patch box and line monitor. In addition, switches or jumpers are usually provided for linking lines on either side of the box. Connection is almost invariably via two 25-way ribbon cables terminated with connectors.

Interface testers

Interface testers are somewhat more complex than simple breakout boxes and generally incorporate facilities for forcing lines into mark or space states, detecting *glitches*, measuring baud rates, and also displaying the format of data words. Such instruments are, not surprisingly, rather expensive.

Oscilloscopes

An oscilloscope can be used to display waveforms of signals present on data lines. It is thus possible to detect the presence of noise and glitches as well as measuring signal voltage levels, and rise and fall times. A compensated ($\times 10$) oscilloscope probe will normally be required in order to minimise distortion caused by test-lead reactance. A digital storage facility can be invaluable when displaying transitory data.

Multimeters

A general-purpose multimeter can be useful when testing static line voltages, cable continuity, terminating resistances, etc. A standard multi-range digital instrument will be adequate for most applications, however, an audible continuity testing range is useful when checking data cables.

Fault finding on RS-232 systems

Fault finding on RS-232 systems usually involves the following basic steps:

(a) ascertain which device is the DTE and which is the DCE. This can usually be accomplished by simply looking at the connectors (DTE equipment is normally fitted with a male connector while DCE equipment is normally fitted with a female connector). Where both devices are configured as DTE (as is often the case) a patch box or null modem should be inserted for correct operation

(b) check that the correct cable has been used. Note that RS-232 cables are provided in a variety of forms; 4-way (for dumb terminals), 9-way (for normal asynchronous data communications), 15-way (for synchronous communications), and 25-way (for universal applications). If in doubt, use a full 25-way cable

(c) check that the same data word format and baud rate has been selected at each end of the serial link

(d) activate the link and investigate the logical state of the data (TXD and RXD) and handshaking (RTS, CTS, etc.) signal lines using a line monitor, breakout box, or interface tester. Lines may be looped back to test each end of the link

(e) if in any doubt, refer to the equipment manufacturer's data in order to ascertain whether any special connections are required and to ensure that the interfaces are truly compatible. Note that some manufacturers have implemented quasi-RS-232 interfaces which make use of TTL signals. These are *not* electrically compatible with the normal RS-232 system

(f) the communications software should be initially configured for the least complex protocol (eg, basic ASCII character transfer without handshaking). When a successful link has been established, more complex protocols may be attempted.

The program listing shows a simple GWBASIC program which can be used to test an asynchronous RS-232 link in full-duplex mode between two PCs (or PC compatibles). Similar programs can be used in other environments or between two quite different machines. The two computers should be linked using a null-modem cable (or null-modem connector) and the program should be entered, saved to disk, and then loaded and run on both computers.

The program can be easily modified to test the COM2 asynchronous port (rather than COM1) by changing the OPEN statement in line 150. This line may also be modified in order to test the link at different baud rates (other than 300 baud) and with different data formats. The OPEN command has the following syntax when used with a communications device:

OPEN "COMn: [speed],[parity],[data],[stop]" AS #filenum

Where:

n	refers to the asynchronous port number (1, 2, 3, etc.)
speed	is the baud rate (150, 300, 600, etc.)
parity	is the parity selected (N = none, E = even, and O = odd)
data	refers to the number of data bits (5, 6, 7 or 8)
stop	refers to the number of stop bits (1, 1.5 or 2)

Readers are advised to consult the appropriate Microsoft GWBASIC or QuickBASIC manuals for further information.

```
100    REM Simple full duplex communications
105    REM test routine using PC COM1 serial port
110    REM Data format; 300 baud, even parity
115    REM seven data bits, one stop bit
120    KEY OFF
130    CLS
140    PRINT "GWBASIC full duplex communications"
150    OPEN "COM1:300, E,7,1" AS #1
160    K$=INKEY$
```

```
170    IF K$=""THEN GOTO 210
180    IF K$=CHR$(3) OR K$=CHR$(27) THEN GOTO 250
190    PRINT #1,K$;
200    PRINT K$;
210    IF EOF(1) THEN GOTO 160
220    C$=INPUT$(LOC(1),#1)
230    PRINT C$;
240    GOTO 160
250    CLOSE #1
260    CLS
270    END
```

Cable modem

Cable modems operate over the ordinary cable-TV network and are connected to the TV outlet at the customer end and the corresponding cable modem termination system (CMTS) at the cable-TV company's end (the Head-End). The cable modem is functionally like a local area network (LAN) interface.

The cable modem is capable of a data rate typically between 3 Mbps and 50 Mbps, and can transmit over a distance of 100 km or more. The CMTS can talk to all the cable modems connected to it, but the cable modems can only talk to the CMTS and not to each other. If two cable modems need to talk to each other, the CMTS will have to relay the messages.

Current systems are based on standards: MCNS/DOCSIS 1.0/1.1 (used in the USA) and DVB/DAVIC 1.3/1.4/1.5 (used in Europe). Cable modems from different vendors will work together, provided that their design is based on the same standard. Version 1.0 of the MCNS standard specified 10 Mbps Ethernet as the only allowable data-interface. By contrast, the DVB/DAVIC standard is totally open and allows any type of interface. Other types of interfaces, including USB, are incorporated in version 1.1 of the MCSN standard, allowing for a wider range of cable modem configurations.

The DOCSIS standard is used in the USA, but is slightly modified to meet European requirements. The European version is called Euro-DOCSIS. Under DOCSIS, the frequency used to transmit data from the cable modem to the CMTS (up-stream) is normally in the 5 MHz to 42 MHz range for USA systems and 5 MHz to 65 MHz for European systems. Data is multiplexed using TDMA (mini-slots in Europe) and modulated on the carrier frequency using QPSK/16-QAM. Data rates are typically 3 Mbps in the up-stream direction.

In the down-stream direction, from the CMTS to the cable modem, transmit frequencies are in the 42 MHz to 850 MHz range for USA systems and 65 MHz to 850 MHz for European systems. Data is multiplexed using TDM in the USA, but MPEG is used in Europe to be compatible with digital video broadcasting (DVB). Data is modulated on a carrier signal using 64/256-QAM modulation. Data rates are typically 27 to 56 Mbps in the down-stream direction.

Most cable-TV networks are hybrid fibre-coax (HFC). The signals are transmitted over fibre-optic cables from the CMTS to a location near the subscriber. At that point, the signal is converted to electrical for transmission over coaxial cables that enter the subscriber premises.

One CMTS will normally drive up to 2000 simultaneous cable modem users on a single TV channel. If more cable modems are required, the number of TV channels has to be increased.

The cable modem can be internal or external. The external cable modem can connect to a number of computers using an ordinary Ethernet connection. Another interface found on external cable modems is the Universal Serial Bus (USB), but this only allows one PC to be connected at any one time. The internal cable modem is usually a PCI bus add-in card that goes inside the PC. This type of cable modem can only be used in desktop PC's and thus may not have galvanic isolation from the mains supply. In some countries, and on some cable-TV networks, it may not be possible to use internal cable modems for technical or regulatory reasons.

The interactive set-top box is also a cable modem and is used in conjunction with a TV set. Its primary function is to provide more TV channels using a limited number of carrier frequencies. This is possible with the use of digital television encoding (DVB). An interactive set-top box provides a return channel, usually a separate telephone line, which gives the user access to the Internet and email using the TV screen as a display.

Data Communications Equipment

ATM media converters

ATM media converters provide a means of interconnecting ATM signals between a variety of different media including shielded twisted pairs, unshielded twisted pairs, coaxial cables, single mode optical fibres and multimode optical fibres. A single ATM media converter can be used to connect two ATM devices operating with dissimilar physical and electrical interfaces. A pair of ATM media converters may be used to connect two devices having the same interface but operating over a data transmission path based on a different physical medium.

ATM rate and media converters

ATM rate and media converters provide asynchronous transfer mode rate conversion between two devices by extracting ATM cells from one interface before sending them over a different interface. A large FIFO (first-in first-out) buffer is used for rate adaptation and the ATM flow control loop is adjusted in accordance with the available bit rate (ABR) in order to avoid cell losses and FIFO overflow. Various interface standards and physical media are usually supported. These may include E1 over coaxial cable and unshielded twisted pair, DS1 over unshielded twisted pair, E3 over coaxial cable, DS3 over coaxial cable, etc.

Baluns

Baluns provide a means of connecting an unbalanced line (eg, coaxial cable) to a balanced line (eg, four-wire twisted pair). Such devices are invariably passive (ie, they require no power).

Baseband modems

Baseband modems (also known as *shorthaul modems*) enable devices such as terminals, computers, controllers, etc. to be interconnected over relatively short distances such as inside buildings, within a site boundary (eg, a college campus), or across a small town. With such a device, a typical range of up to about 10 km can be achieved at a data rate of 9.6 kbps using a conventional two-wire telephone line.

Bridges

A bridge is a device that interconnects two or more networks of the same type (eg, two networks based on the Ethernet standard or two Token Ring IEEE 802.5 LANs). Bridges operate within the Data Link Layer of the ISO model for OSI. *Adaptive bridges* are able to configure themselves by constructing a table of user addresses. *Remote bridges* provide a means of interconnecting two networks in different locations (eg, a central office network with a remote office network). In this case, each network is fitted with a bridge and these are then linked together with a digital circuit (eg, an E1 line).

Coaxial multiplexers

Coaxial multiplexers enable two (or more) coaxial cables to be connected to an incoming or outgoing coaxial cable. The same impedance is presented at each port and no power is required by the device.

Current loop converters

Current loop converters convert standard RS-232/V.24 signals into bi-directional current loop signals (either 20 or 60 mA). Data rates of up to 19.2 kbps and distances of up to 5 or 6 km can be supported by such a device. Active current loop converters provide a source of loop current while passive current loop converters accept the current supplied from another current loop device.

Digital service access devices

Digital service access devices provide the digital interface for the customer premises equipment (CPE) to the carrier's digital services. Digital services access devices extend the network to the customer's premises and, in this respect, they differ from conventional digital modems.

Fibre optic modems

Standard fibre optic modems are designed to operate with optical fibres rather than copper cables. Data rates of 128, 256, 384, 512, 768, 1024, 1544 (T1) and 2048 kbps (E1) can be achieved with such devices. Typical distances are up to 5 km (850 nm multimode fibre), 20 km (1300 nm single mode fibre) and 50 km (laser diode). The digital interface usually supports one or more of the following standards; V.24, V.35, X.21, RS-520 and G.703. High-speed fibre optic modems support the use of data rates of 34.368 (E3), 44.736 (T3) and 51.84 Mbps (STS-1).

Data rates of 10 Gbps are transmitted over fibre networks, using WDM and externally modulated lasers. The terminal equipment is rack mounted and cannot really be described as a 'modem'.

Frame relays (packet assemblers/disassemblers)

Frame relays can be used to connect a number of asynchronous data channels to an X.25 or frame relay network. Such devices can also serve as access units to an X.25 private or public network or as access servers to a mainframe with an X.25 port.

Gateways

Gateways perform the functions of both routing and media conversion (eg, converting circuit-oriented or analogue information, such as voice into TCP/IP packets, and vice versa). One example of their use is in local telephone exchanges, to provide access to the Internet. In this situation, the gateway off-loads IP traffic from the PSTN as soon as possible. Moving IP traffic off the PSTN frees up the capacity of switches and trunks to allow them to handle circuit-switched calls. New sessions to ISPs are established by linking the dialled number with an IP address, so that one gateway can support circuit-switched access to multiple ISPs, thereby controlling ISP and network operator cost.

Hubs

Hubs help to simplify the wiring of a LAN by providing a common physical point for cabling. Hubs do not affect the way in which a network operates – once inside the hub, the original bus or ring topology is preserved. A typical Ethernet hub provides sixteen 10Base T ports within one physical unit. Two further ports are available for expansion by 'stacking'.

Interface converters

Interface converters provide a means of linking two dissimilar networks including coping with different physical connections, signal characteristics, and different meanings of exchanged signals. Interface converters are often required to also cope with a change of data rate in which case they are more properly known as 'rate and interface converters'.

Inverse multiplexers

Inverse multiplexing is used for the transmission of a high-speed data channel over two or more lower speed WAN links. As an example, an aggregate data rate of up to 384 kbps (V.35) can be realised using up to six multiplexed leased lines at 64 kbps. Inverse multiplexers must be present at both sites – the remote multiplexer reconstructs the high-speed data signal from the signals present on each of the lower speed links. Inverse multiplexers can also be used to back up high-speed leased/digital lines using one or more ISDN links which can be brought into service during times of peak traffic demand.

LAN managers

LAN managers comprise software and hardware (usually based on a PC platform) which will provide direct on-line supervision of network configuration, diagnostics, monitoring and control. Most currently available LAN management software runs under Microsoft Windows, NT or Windows 95. In order to determine network loading and identify problem areas, real-time statistical information can be provided displayed in various formats, including line graph, bar chart or tabular form. More sophisticated LAN managers feature automatic recognition of access units, enhanced diagnostic and security features, and an ability to control third party equipment.

Line termination units

Line termination units provide signal conditioning (equalisation and adaptive filtering) to combat the effects of attenuation, distortion and noise present in lines. At distances of up to about 5 km with conventional four-wire copper cables, they can both eliminate the need for repeaters and provide an effective alternative to the use of fibre optical cables. In some cases, an embedded channel may be provided for control and diagnostics.

Medium range ('voice band') modems

Standard modems operating with data rates of up to 19.2 kbps can be used with conditioned lines at distances of up to 100 km (distances of up to about 60 km can be achieved when conditioned lines are not available). Such modems operate in a similar manner to conventional telephone-line modems but cater for both synchronous and asynchronous operation and generally incorporate diagnostics to V.54 with built-in bit error rate testers.

Multiplexers

Multiplexers provide an efficient and cost-effective method of integrating data, voice, fax and LAN traffic over digital data services, leased lines, and ISDN. Modular multiplexer design allows services such as V.35, RS-530, V.24/RS-232 and X.21 to share the same leased line or private channel at data rates typically in the range 9.6 to 768 kbps.

Packet switching access units

Packet switching access units provide encapsulation of protocols over a Frame Relay or an X.25 network. Various protocols can usually be accommodated including X.25, Frame Relay, STEM and HDLC.

Rate converters

A typical application for a rate converter is that of facilitating the connection of a 56 or 48 kbps terminal to a 64 or 56 kbps line. A facility for a synchronous sub-channel at 4.8 or 9.6 kbps may also be provided.

Rate and interface converters

Combined rate and interface converters are also available to cope with situations in which the physical interconnections, signal interpretation and data rates may differ on both sides of the interface. A typical application for such a device would be that of converting an E1 frame (2.048 Mbps) into two T1 frames (1.024 Mbps) – hence allowing E1 equipment to operate with only T1 facilities.

Repeaters

Repeaters amplify and regularise the voltage and/or current levels of a digital signal. This enables them to compensate both for losses in the line and distortion due to non-linearity of the line's frequency response characteristic. The content of the digital signal (in terms of the data present and its rate) remain unchanged by a repeater. Repeaters can be line-powered (using either through or loop modes) or may derive their power supply from a battery or standard a.c. mains supply outlet. In many cases, local loopback support may be available so that the repeater – as well as the line up to that point in the circuit – can be tested. A typical E1 repeater (2.048 Mbps) will provide satisfactory operation with up to 2 km of conventional cable and line attenuation of up to 40 dB. Beyond that distance, repeaters can be chained to achieve greater distances.

Routers

Routers, like bridges, can be used to interconnect LANs. Unlike bridges, routers operate at the Network Layer of the ISO model for OSI as well as the Data Link Layer. This permits the use of higher level addresses. Each router has its own network address and only needs the address of another router in order to reach all of the users connected to the network served by that router.

Switches

Ethernet switches increase network performance by not passing any unnecessary traffic onto individual network segments attached to the switch. They also filter packets a bit like a router does. When a packet arrives, the header is checked to determine which segment the packet is destined for, and then the switch forwards the packet to that segment. This prevents the packet being forwarded onto unnecessary segments, thus reducing the traffic. To reduce the switch workload, nodes that inter-communicate frequently should be placed on the same segment. Switches work at the MAC layer level.

Cut-through switches

Cut-through switches use either a cross-bar or cell back-plane architecture. Only the first few bytes of the packet are read to obtain the source and destination addresses. The packets are then passed through to the destination segment without checking the rest of the packet for errors. The lack of checking means that invalid packets can still be passed onto other segments, but ensures that there is little throughput delay.

Cross-bar switches read the destination address then immediately forward the packet. Although it acts as a simple repeater once the path is established, it can introduce delay if the destination port is busy because it may need to buffer the packet.

Cell back-plane switches break the frame into small fixed cell lengths. Each cell is labelled with special headers that contain the address(s) of the destination port. The cells are buffered at the destination port and then re-assembled into a packet. The packet is then transmitted onward. The data rate on the back-plane is significantly greater than the aggregate data rate of the ports. In overloaded networks, cell back-plane switching gives a better performance than cross-bar switching

Store and forward switches

Store and forward switches examine the entire packet. Each packet is buffered at the switch input and then examined. The switch removes any bad packets that it detects and good packets are forwarded to the correct segment. Store and forward switches detect more errors than cut-through switches, although they do impose a small throughput delay.

124

Use of rate and interface converters to enable T1 equipment to make use of E1 transmission media

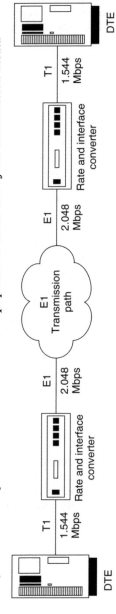

Use of short range modems to link LAN on two sites up to 4km apart

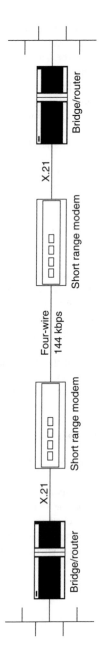

Use of short range modems to link computers on two sites up to 4 km apart

Use of short range modems to link a computer with a remote terminal or PC at up to 10 km

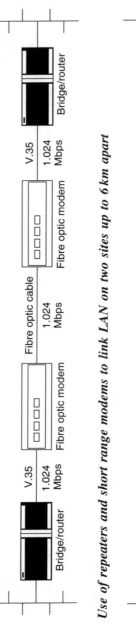

Use of fibre optic modems to link LAN on two sites up to 50 km apart

Use of repeaters and short range modems to link LAN on two sites up to 6 km apart

Use of inverse multiplexers to increase data transfer rate with leased lines

Use of inverse multiplexers to increase data transfer rate over multiple E1 links

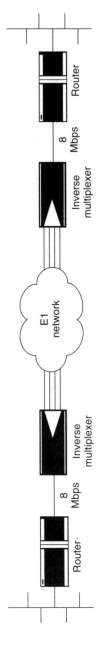

Use of inverse multiplexers to provide leased line backup

Use of multiplexers to provide voice/LAN communications via a high-speed WAN

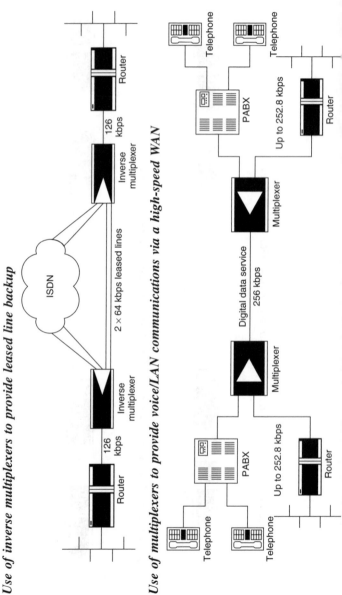

Use of multiplexers to provide integrated voice/fax/LAN/data communications

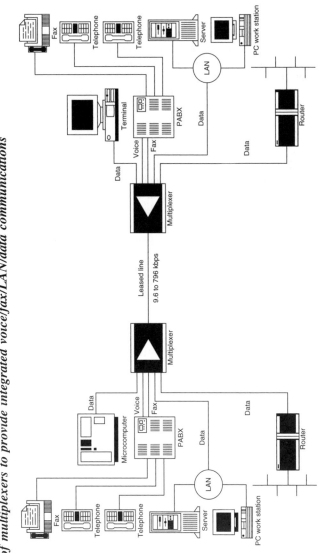

6 Parallel interfaces

The advantage of a parallel interface is the speed at which data can be transferred from a computer to a peripheral device. Usually, parallel interfaces have eight data-carrying connections, so the data can be transferred at least eight times faster than an equivalent serial connection. An alternative way of looking at this is that the data rate on each data line to a peripheral can be eight or more times slower in order to carry the same amount of data in a given period.

The most basic parallel connection a computer bus, these can be 8, 16 or 32 bits wide. A microprocessor can read from or write to this bus. Integrated circuits acts as buffers so that external data is only allowed onto the bus when the microprocessor is in its read mode. The buffer is arranged to output data when the microprocessor is in its write mode.

There are two common parallel interfaces used external to the computer: the Centronics printer port and the IEEE-488 bus. These are described later in this chapter.

Parallel I/O devices

Parallel I/O devices allow a byte of data to be transferred at a time between computer systems and external devices. Parallel I/O is relatively easy to implement since it only requires an arrangement based on 8-bit buffers or latches. The software and hardware requirements of this form of I/O are thus minimal.

Parallel I/O devices enjoy a variety of names depending upon their manufacturer. Despite this, parallel I/O devices are remarkably similar in internal architecture and operation with only a few minor differences distinguishing one device from the next.

Programmable parallel I/O devices can normally be configured (under software control) in one of several modes:

(a) all eight lines configured as inputs
(b) all eight lines configured as outputs
(c) lines individually configured as inputs or outputs.

In addition, extra lines to I/O lines are normally available to facilitate *handshaking*. This provides a means of controlling the exchange of data between a computer system and external hardware. The nomenclature used for parallel I/O lines and their function tends to vary from chip to chip. The following applies to the majority of devices:

PA0 to PA7   Port A I/O lines; 0 corresponds to the least significant bit (LSB) whilst 7 corresponds to the most significant bit (MSB)

CA1 to CA2   Handshaking lines for Port A; CA1 is an interrupt input whilst CA2 can be used as both an interrupt input and peripheral control output

PB0 to PB7   Port B I/O lines; 0 corresponds to the least significant bit (LSB) whilst 7 corresponds to the most significant bit (MSB)

CB1 to CB2   Handshaking lines for Port B; CB1 is an interrupt input whilst CB2 can be used as both an interrupt input and peripheral control output.

Programmable I/O devices are invariably TTL-compatible and buffered to support at least one conventional TTL load. Several programmable parallel I/O devices have port output lines (usually the B group) which are able to source sufficient current to permit direct connection to the base of a conventional or Darlington-type transistor. This device can then be used as a relay or lamp driver. Alternatively, high-voltage open-collector octal drivers may be connected directly to the port output lines.

Internal architecture of a representative parallel I/O device

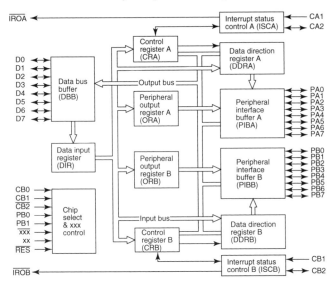

Internal registers of a typical programmable parallel I/O device

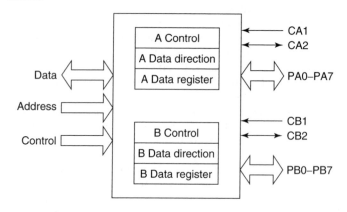

CPU interface to a programmable parallel I/O device

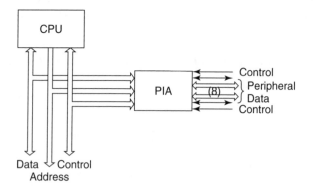

Centronics printer interface

The Centronics interface has established itself as the standard for parallel data transfer between a microcomputer and a printer. The standard is based on 36-way Amphenol connector (part no: 57–30360) and is suitable for distances of up to 2 m.

Parallel data is transferred into the printer's internal buffer when a strobe pulse is sent. Handshaking is accomplished by means of acknowledge (ACKNLG) and busy (BUSY) signals.

Centronic printer interface pin assignment

Pin No.	Abbreviation	Signal\function
1	STROBE	Strobe (active low to read data)
2	DATA 1	Data line 1
3	DATA 2	Data line 2
4	DATA 3	Data line 3
5	DATA 4	Data line 4
6	DATA 5	Data line 5
7	DATA 6	Data line 6
8	DATA 7	Data line 7
9	DATA 8	Data line 8
10	ACKNLG	Acknowledge (pulsed low to indicate that data has been received)
11	BUSY	Busy – taken high under the following conditions:
		(a) during data entry
		(b) during a printing operation
		(c) when the printer is OFF-LINE
		(d) during print error status
12	PE	Paper end (taken high to indicate that the printer is out of paper)
13	SLCT	Select (taken high to indicate that the printer is in the selected state)
14	AUTO FEED XT	Automatic feed (when this input is taken low, the printer is instructed to produce an automatic line feed after printing. This function can be selected internally by means of a DIP switch)
15	n.c.	Not connected (unused)
16	0 V	Logic ground
17	CHASSIS GND	Printer chassis (normally isolated from logic ground at the printer)
18	n.c.	Not connected (unused)
19 to 30	GND	Signal ground (originally defined as 'twisted pair earth returns' for pin numbers 1 to 12 inclusive)
31	INIT	Initialise (this line is pulsed low to reset the printer controller)
32	ERROR	Error – taken low by the printer to indicate:
		(a) PAPER END state
		(b) OFF-LINE state
		(c) error state
33	GND	Signal ground
34	n.c.	Not connected (unused)
35	LOGIC 1	Logic 1 (usually pulled high via 3.3 kohm)
36	SLCT IN	Select input (data entry to the printer is only possible when this line is taken low, but this function may be disabled by means of an internal DIP switch)

Notes:

1. Signals, pin numbers, and signal directions apply to the printer
2. Alternative types of connector (such as the 25-way D type, PCB edge, etc.) are commonly used at the microcomputer
3. All signals are standard TTL levels
4. ERROR and ACKNLG signals are not supported on some interfaces

Centronics interface pin connections

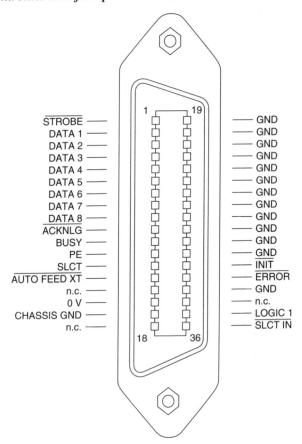

IEEE-488 interface standard

The IEEE-488 bus (also known as the Hewlett Packard instrument bus and the general-purpose instrument bus) provides a means of inter-connecting a microcomputer controller with a vast range of test and measuring instruments. The bus is ideally suited to the implementation of automatic test equipment (ATE), and it has become increasingly popular in the last decade with a myriad of applications which range from routine production tests to the solution of highly complex and specialised measurement problems.

In the past, IEEE-488 facilities have tended to be available within only the more expensive test equipment. The necessary interface is, however, becoming increasingly commonplace in medium- and low-priced instruments. This trend reflects not only an increased demand from the test equipment user, but also the availability of low-cost dedicated IEEE-488 controller chips.

Nowadays, most items of modern electronic test equipment (such as digital voltmeters and signal generators) and many items of peripheral equipment are either fitted with the necessary IEEE-488 interface as standard or can be upgraded with optional IEEE-488 interface cards. This provision allows them to be connected to a microcomputer controller via the IEEE-488 bus so that the controller can be used both to supervise their operation and process the data which they collect.

IEEE-488 devices

The IEEE-488 standard provides for the following categories of device:

(a) Listeners

Listeners can receive data and control signals from other devices connected to the bus, but are not capable of generating data. An obvious example of a listener is a signal generator.

(b) Talkers

Talkers are only capable of placing data on the bus and cannot receive data. Typical examples of talkers are magnetic tape, magnetic stripe, and bar code readers. Not that, while only one talker can be active (ie, presenting data to the bus) at a given time, it is possible for a number of listeners to be active simultaneously (ie, receiving and/or processing the data).

(c) Talkers and listeners

The function of a talker and listener can be combined in a single instrument. Such instruments can both send data to and receive data from the bus. A digital multimeter is a typical example of a talker and listener. Data is sent to it in order to change ranges and is returned to the bus in the form of digitised readings of voltage, current, and resistance.

(d) Controllers

Controllers are used to supervise the flow of data on the bus and provide processing facilities. The controller within an IEEE-488 system is invariably a microcomputer and, whilst some manufacturers provide dedicated microprocessor based IEEE-488 controllers, this function is often provided by means of a PC or PC-compatible microcomputer.

IEEE-488 bus signals

The IEEE-488 bus uses eight multi-purpose bi-directional parallel data lines. These are used to transfer data, addresses, commands, and status bytes. In addition, five bus management and three handshake lines are provided.

The connector used for the IEEE-488 bus is invariably a 24-pin Amphenol type having the following pin assignment:

Pin number	Abbreviation	Function
1	DIO1	Data line 1
2	DIO2	Data line 2
3	DIO3	Data line 3
4	DIO4	Data line 4
5	EOI	End or identify. This signal is generated by a talker to indicate the last byte of data in a multi-byte data transfer. EOI is also issued by the active controller to perform a parallel poll by simultaneously asserting EOI and ATN.
6	DAV	Data valid. This signal is asserted by a talker to indicate that valid data has been placed on the bus.
7	NRFD	Not ready for data. This signal is asserted by a listener to indicate that it is not yet ready to accept data.
8	NDAC	No data accepted. This signal is asserted by a listener whilst data is being accepted. When several devices are simultaneously listening, each device releases this line at its own rate (the slowest device will be the last to release the line).
9	IFC	Interface clear. Asserted by the controller in order to initialise the system in a known state.
10	SRQ	Service request. This signal is asserted by a device wishing to gain the attention of the controller. Note that this line employs wire-OR'd logic.
11	ATN	Attention. Asserted by the controller when placing a command on to the bus. When the line is asserted this indicates that the information placed by the controller on the data lines is to be interpreted as a command. When it is not asserted, information placed on the data lines by the controller must be interpreted as data. ATN is always driven by the active controller.
12	SHIELD	Shield.
13	DIO5	Data line 5

Pin number	Abbreviation	Function
14	DIO6	Data line 6
15	DIO7	Data line 7
16	DIO58	Data line 8
17	REN	Remote enable. This line is used to enable or disable bus control (thus permitting an instrument to be controlled from its own front panel rather than from the bus).
18–24	GND	Ground/common signal return.

Notes:

1. Handshake signals (DAV, NRFD, and NDAC) employ active low open-collector outputs which may be used in a wired-OR configuration.
2. All remaining signals are fully TTL compatible and are active low.

IEEE-488 bus connector

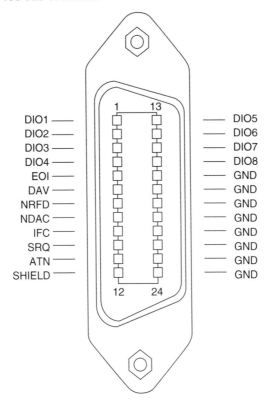

IEEE-488 commands

Bus commands are signalled by taking the ATN line low. Commands are then placed on the bus by the controller and directed to individual devices by placing a unique address on the lower five data bus lines. Alternatively, universal commands may be issued to all of the participating devices.

Handshaking

The IEEE-488 bus uses three handshake lines (DAV, NRFD, and NDAC). The handshake protocol adopted ensures that reliable data transfer occurs at a rate determined by the *slowest* listener.

A talker wishing to place data on the bus first ensures that NDAC is in a released state. This indicates that all of the listeners have accepted the previous data byte. The talker then places the byte on the bus and waits until NRFD is released. This indicates that all of the addressed listeners are ready to accept the data. Finally, the talker asserts DAV to indicate that the data on the bus is valid.

IEEE-488 handshake sequence

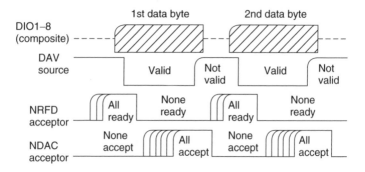

Service requests

The service request (SRQ) line is asserted whenever a device wishes to attract the attention of the active controller. SRQ essentially behaves as a shared interrupt line since all devices have common access to it. In order to determine which device has generated a service request, it is necessary for the controller to carry out a poll of the devices present. The polling process may be carried out either serially or in parallel.

In the case of *serial polling*, each device will respond to the controller by placing a status byte on the bus. DIO7 will be set if the device in question is requesting service, otherwise this data bit will be reset. The active controller continues to poll each device present in order to determine which one has generated the service request. The remaining bits within the status byte are used to indicate the status of a device and, once the controller has located the device which requires service, it is a fairly simple matter to determine its status and instigate the appropriate action.

In the case of *parallel poling*, each device asserts an individual data line. The controller can thus very quickly determine which device requires attention. The controller, however, cannot ascertain the status of the device which has generated the service request at the same time. In some cases it will be necessary, therefore, to carry out a subsequent serial poll of the same device in order to determine its status.

Multiline commands

The controller sends multiline commands over the bus as data bytes with ATN asserted. Multiline commands are divided into five groups, as follows:

Command group	Abbreviation	Function	Command byte
Addressed command	ACG	Used to select bus functions affecting listeners (eg, GTL which restores local front-panel control of an instrument)	00–0F
Universal command	UCG	Used to select bus functions which apply to all devices (eg, SPE which instructs all devices to output their serial poll status byte when they become the active talker)	10–1F
Listen address	LAG	Sets a specified device to listen	20–3E
	UNL	Sets all devices to unlisten status	3F
Talk address	TAG	Sets a specified device to talk	40–5E
	UNL	Sets all devices to untalk status	5F
Secondary command	SCG	Used to specify a device sub-address or sub-function (also used in a parallel poll configure sequence)	60–7F

IEEE-488 command codes

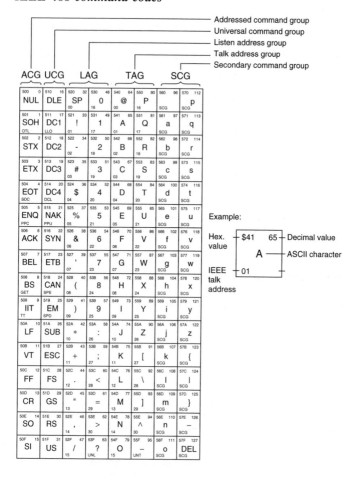

IEEE-488 bus configuration

Since the physical distance between devices is usually quite small (less than 20 m), data rates may be relatively fast. Data rates of between 50 kbytes s^{-1} and 250 kbytes s^{-1} are typical: however, to cater for variations in speed of response, the slowest listener governs the speed at which data transfer takes place.

Typical IEEE-488 bus configuration

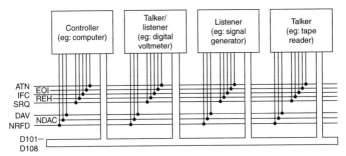

IEEE-488 software

In order to make use of an IEEE-488 bus interface, it is necessary to have a DOS resident driver to simplify the task of interfacing with control software. The requisite driver is invariably supplied with the interface hardware (ie, the IEEE-488 expansion card). The driver is installed as part of the normal system initialisation and configuration routine and, thereafter, will provide a software interface to applications packages or bespoke software written in a variety of languages (eg, BASIC, Pascal, and C). The user and/or programmer is then able to access the facilities offered by the IEEE-488 bus using high-level IEEE-488 commands such as REMOTE, LOCAL, ENTER, OUTPUT, etc.

The following are a typical set of high-level language commands which may be used to program an IEEE-488 system:

Command	Function
ABORT	Terminate the current selected device and command. If no device is given, the bus is cleared and set to the state given in the last CONFIG command eg **ABORT 1** terminates device 1
CLEAR	Clear or reset the selected devices or all devices. If no device is given, the bus is cleared and set to the state given in the last CONFIG command eg **CLEAR 10** resets device 10

Command	Function
CONFIG	Configures the bus to a given set of requirements. The bus will remain in the configured state until it is reconfigured eg **CONFIG TALK = 2 LISTEN = 1, 3, 4** configures device number 2 as a talker and devices 1, 3, and 4 as listeners
ENTER	Enters bus data from the selected device into a specified string array (the array must have been previously dimensioned). A flag (FLAG%) will contain any error codes returned eg **ENTER 10[$, 0, 15]** enters data from address 10, array elements 0 to 18
EOI	Sends a data byte to the selected device with EOI asserted. The bus must have been programmed to talk before the command is executed. The variable contains the data to be transferred eg **EOI 12[$]** issues an EOI with the last byte of the string to listener 12
LOCAL	Sets the selected device(s) to the local state. If no device is specified then all devices on the bus are set to local eg **LOCAL 10, 11** sets devices 10 and 11 to local state
LOCKOUT	Locks out (on a local basis) the specified device(s). The devices cannot be set to local except by the bus controller eg **LOCKOUT 9, 10** performs a local lockout on devices 9 and 10
OUTPUT	Outputs a string to the selected listener(s). If no listener is specified in the command then all listeners will receive the specified string eg **OUTPUT 9, 11 [$E]** outputs the specified string using even parity
PARPOL	Reads the status byte for the devices which have been set for parallel polling eg **PARPOL** reads status byte from a parallel polled device
PASCTL	Passes control of the bus to the specified device. Thereafter, the issuing PC controller will adopt the role of talker/listener eg **PASCTL 5** passes control of the bus to device 5 (which must be a bus controller)
PPCONF	Sets the parallel polling configuration for the specified device eg **PPCONF 12** selects parallel polling for device 12
PPUNCF	Resets the parallel polling configuration for the specified device eg **PPUNCF 12** de-selects parallel polling for device 12
REMOTE	Selects remote operation for the specified device(s) eg **REMOTE 9, 10, 11** selects remote operation for devices 9, 10 and 11

Command	Function
REQUEST	Requests service from an active bus controller (used only when the computer itself is the current bus controller) eg **REQUEST**　requests service from the current bus controller
STATUS	Reads a (serial polled) status byte from the selected device eg **STATUS 8**　reads the status byte (serial polled) from device 8
SYSCON	Configures the system for a particular user configuration. The command initialises a number of system variables including: MAD　the address of the system controller CIC　the controller board in charge (more than one IEEE-488 bus controller board may be fitted to a computer) NOB　the number of IEEE-488 bus controller boards fitted (1 or 2) BA0　the base I/O address for the first bus controller board (ie, board 1) BA1　the base I/O address for the second bus controller board (ie, board 2) eg **SYSCON MAD = 3, CIC = 1NOB = 1, BA0 = &H300** Configures the system as follows: Computer bus controller address = 3 Controller board in charge = 1 Number of boards fitted = 1 Base address of the controller board = 300 hex
TIMEOUT	Sets the timeout duration when transferring data to and from devices. An integer number (eg, VAR%) in the range 0 to 65000 is used to specify the time. For a standard IBM-PC/XT the time (in seconds) is equivalent to 3.5*VAR% whilst for an IBM-PC/AT the time is approximately 1.5*VAR%
TRIGGER	Sends a trigger message to the selected device (or group of devices) eg **TRIGGER 9, 10**　triggers devices 9 and 10

Note: If a command is issued by a device which is not the current controller then an error condition will exist

IEEE-488 programming

Programming an IEEE-488 system is relatively straightforward and it is often possible to pass all control information to the DOS resident software driver in the form of an ASCII encoded string. The command string is typically followed by three further parameters:

(a) the variable to be used for output or input (either an integer number or a string)
(b) a flag (integer number) which contains the status of the data transaction (eg, an error or transfer message code)

(c) the address of the interface board (either 0 or 1 or the physical I/O base address)

A command is executed by means of a CALL to the relevant DOS interrupt. The syntax of an interpreted BASIC (BASIC-A or GWBASIC) statement would thus be:

```
CALL IEEE(CMD$, VAR$, FLAG%, BRD%)
```

where:

IEEE	is the DOS interrupt number
CMD$	is the ASCII command string
VAR$	is the variable to be passed (where numeric data is to be passed, VAR$ is replaced by VAR%)
FLAG%	is the status or error code, and
BRD%	is the board number (0 or 1)

As an example, the following GWBASIC code configures a system and then receives data from device 10, printing the value received on the screen:

```
100  REM System configuration
110  DEF SEG=&H2000
120  BLOAD "GPIBBASI.BIN", 0
130  IEEE=0
140  FLAG%=0
150  BRD%=&H300
160  CMD$="SYSCON MAD=3, CIC=1, NOB=1, BA0=768"
170  CALL IEEE(CMD$, A$, FLAG%, BRD%)
180  PRINT "System configuration status: "; HEX$(FLAG%)
200  REM Get string data from device 10
210  B$=SPACE$(18)
220  CMD$="REMOTE 10"
230  CALL IEEE(CMD$, B$, FLAG%, BRD%)
240  PRINT "Remote device 10 return flag: "; HEX$(FLAG%)
250  CMD$="ENTER 10[$, 0, 17]"
260  CALL IEEE(CMD$, B$, FLAG%, BRD%)
270  PRINT "Enter from device 10 return flag: "; HEX$(FLAG%)
280  PRINT "Data received from device 10: "; B$
290  END
```

Line 110 defines the start address of a block of RAM into which the low-level interrupt code is loaded from the binary file GPIBBASI.BIN (line 120). The IEEE interrupt number (0) is allocated in line 130 whilst the message/status code is initialised in line 140. Line 150 selects the base I/O address (in this example, for the Metrabyte MBC-488 board) and the system configuration command string is defined in line 160 (note that the PC bus controller is given address 3 and a single IEEE-488 bus interface board is present).

The status flag (returned after configuring the system by means of the CALL made in line 170) is displayed on the screen in hexadecimal format (line 180). An empty string (B$) is initialised in line 210 (this will later receive the data return from device 10). Device 10 is selected as the remote device in lines 220 and 230 whilst line 240 prints the returned status flag for this operation. Data is then read from device 10 (lines 250 and 260) and, finally, the status code and returned data are displayed in lines 270 and 280.

In most cases, it will not be necessary to display returned status codes. However, it is usually necessary to check these codes in order to ascertain whether a particular bus transaction has been successful and that no errors have occurred. Furthermore, a more modern BASIC (eg, Microsoft QuickBASIC) will allow programmers to develop a more structured approach to controlling the IEEE-488 interface with command definitions, error checks, and CALLs consigned to sub-programs.

7 Communication protocols

Communication protocols are the sets of rules and formats necessary for the effective exchange of information within a data communication system. The three elements of a communication protocol are *syntax* (data format, coding, and signal level definitions), *semantics* (synchronisation, control, and error handling), and *timing* (sequencing of data and choice of data rate).

Communication protocols must exist on a range of levels, from the physical interconnection at one extreme to the application responsible for generating and processing the data at the other. It is useful, therefore, to think of protocols as *layered*, with each layer interacting with the layers above and below. This is an important concept and one which leads directly to the ISO seven-layered model for OSI.

ISO model for open systems interconnection

The International Standards Organisation (ISO) model for open systems interconnection (OSI) has become widely accepted as defining the seven layers of protocol which constitute a communication system.

1. Physical layer

The physical layer describes the physical circuits which provide a means of transmitting information between two users. The physical layer is concerned with such items as line voltage levels and pin connections

2. Data link layer

The data link layer defines protocols for transferring messages between the host and network and vice versa. The layer is also responsible for flow control, error detection and link management

3. Network layer

The network layer supports network connections and routing between two hosts and allows multiplexing of several channels via a common physical connection

4. *Transport layer*

The transport layer provides for the transparent transfer of data between end systems which might, for example, organise data differently

5. *Session layer*

The session layer supports the establishment, control and termination of dialogues between application processes. The layer facilitates full duplex operation and maintains continuity of session connections. It also supports synchronisation between users' equipment and generally manages the exchange of data

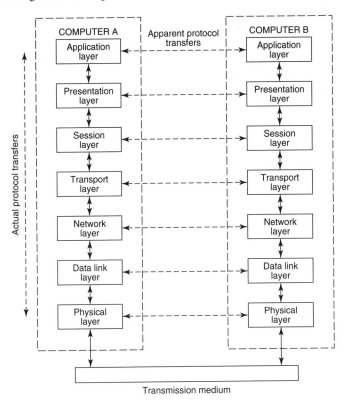

6. Presentation layer

The presentation layer resolves the differences in representation of information used by the application task so that each task can communicate without knowing the representation of information used by a different task (eg, different data syntax)

7. Application layer

The application layer is the ultimate source and sink for data exchange. It provides the actual user information processing function and application specific services by translating user requests into specific network functions

NB: Layers 1 to 3 of the ISO model are often referred to as *communication-oriented* layers. Layers 5 to 7, on the other hand, are referred to as *application-oriented*. In this context, the fourth layer can be thought of as a bridge between the communication and application-oriented layers of the ISO model.

8 Local area networks

A local area network (LAN) is a network which covers a limited area and which generally provides a high data rate capability. A LAN is invariably confined to a single site (ie, a building or group of buildings) and provides for the exchange of information and efficient use of shared resources within the site.

In general a LAN should:

- conform to a well defined international standard supported by a number of manufacturers and vendors
- support a high data rate (typically 1 to 10 Mbps)
- have a maximum range of typically at least 500 metres and, in some cases, as much as 10 km
- be capable of supporting a variety of hardware independent devices (connected as *nodes*)
- provide high standards of reliability and data integrity
- exhibit minimal reliance on centralised components and controlling elements
- maintain performance under conditions of high loading
- allow easy installation and expansion
- readily permit maintenance, reconfiguration, and expansion.

LAN topology

Local area networks are often categorised in terms of the topology which they employ. The following topologies are commonly encountered; *star, ring, tree,* and *bus* (the latter is a tree which has only one *trunk* and no *branches*).

In star topology, a central switching element is used to connect all of the needs within the network. A node wishing to transmit data to another node must initiate a request to the central switching element which will then provide a dedicated path between them, once the circuit has been established, the two nodes may communicate as if they were connected by a dedicated point-to-point path.

Ring topology is characterised by a closed loop to which each node is attached by means of a repeating element. Data circulates around the ring on a series of point-to-point links which exist between the repeaters. A node wishing to transmit must wait for its turn and then send data onto the ring in the form of a *packet* which must contain both the source and destination addresses as well as the data itself.

Upon arrival at the destination node, the data is copied into a local buffer. The packet continues to circulate until it returns to the source node, hence providing a form of acknowledgement.

Bus and tree topologies both employ a multiple-access broadcast medium and hence only one device can transmit at any time. As with ring topology, transmission involves the generation of a packet containing source and destination address field together with data.

Star LAN topology

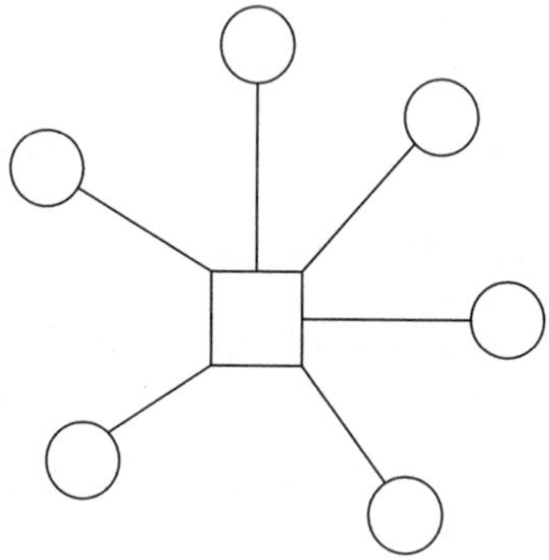

Ring LAN topology

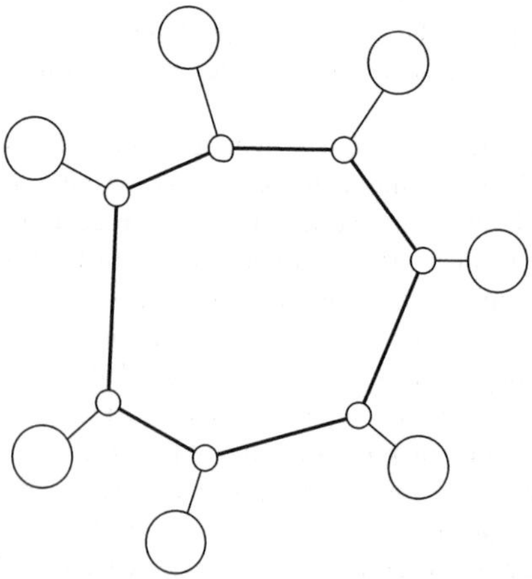

Bus LAN topology

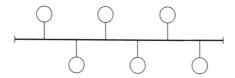

Tree LAN topology

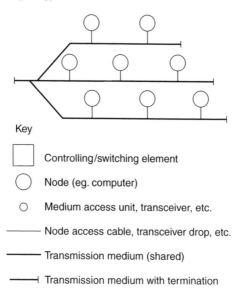

Key

☐ Controlling/switching element

◯ Node (eg. computer)

○ Medium access unit, transceiver, etc.

──── Node access cable, transceiver drop, etc.

━━━ Transmission medium (shared)

━━┤ Transmission medium with termination

Broadband and baseband transmission

Local area networks are available which support either *broadband* or *baseband* transmission. In the former case, information is modulated onto a radio frequency carrier which is passed through the transmission medium (eg, coaxial cable). In the latter case, digital information is passed directly through the transmission medium. It is important to note that broadband LANs can exploit frequency division multiplexing which allows a number of modulated radio frequency carriers (each with its own digital signal) to be simultaneously present within the transmission medium. Baseband LANs can only support one information signal at a time within the transmission medium.

IEEE 802 standards

The IEEE Local Network Standards Committee has developed a series of standards for local area networks. These standards have been produced with reference to the ISO model for OSI and they are summarised here:

General management, addressing and internetworking

| IEEE 802.1 (Part A) | Overview and architecture. |
| IEEE 802.1 (Part B) | Addressing, internetworking, and network management. |

Logical link control

IEEE 802.2   Logical link control (LLC) employed in conjunction with the four media access standards defined under IEEE 802.3, 802.4, 802.5, and 802.g.

Media access control

IEEE 802.3   Carrier sense multiple access and collision detection (CSMA/CD) access method and physical layer specifications.
Note: The European Computer Manufacturers' Association (ECMA) has produced a set of standards which bears a close relationship to that of IEEE 802.3. ECMA standards 80, 81 and 82 relate to CSMA/CD baseband LAN coaxial cables, physical layer, and link layer respectively

IEEE 802.4   Token-passing bus access method and physical layer specifications.

IEEE 802.5   Token-passing ring access method and physical layer specifications.

IEEE 802.6   Metropolitan network access method and physical layer specifications.

IEEE 802.7   Recommended Practices for Broadband Local Area Networks.

IEEE 802.8   Recommended Practice for Fiber Optic LAN/MAN Networks.

IEEE 802.9   Integrated Services (IS) LAN Interface.

IEEE 802.10   Interoperable LAN/MAN Security (SILS).

IEEE 802.11   Wireless LAN Specifications (2.4 GHz and 5 GHz).

IEEE 802.12   Demand Priority Access Method: for 100 Mb/s operation.

IEEE 802.14   Cable-TV access method and physical layer specification.

Relationship between IEEE 802 standards and the ISO model

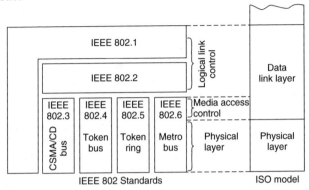

Typical LAN selection flowchart

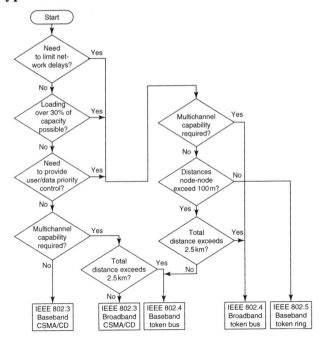

Popular network standards

Ethernet

The Ethernet standard follows IEEE 802.3 and has been widely accepted by a number of independent systems suppliers. Ethernet's popularity stems from a number of factors including the availability of VLSI controllers which permit cost-effective implementation of a network and the high data rate of 10 Mbps.

The Ethernet link control layer employs carrier sense multiple access with collision detection (CSMA/CD) and an HDLC-type frame structure is employed. Basic network physical layer components comprise coaxial cables for transmission media (maximum segment length 500 metres), transceivers with collision detection circuitry, a transceiver drop cable (maximum length 50 metres) which connects data terminal equipment to a nearby transceiver unit, and a controller board. This latter device is responsible for frame assembly/disassembly, handling source and destination addressing, detection of physical transmission errors, collision detection and retransmission.

Where network distances are to exceed 500 metres, multiple segments are employed and these are linked by means of repeaters. Point-to-point links are used to link together segments which are separated by physical distances of up to 1 km. Such links act as a repeater divided into two sections.

Ethernet packet format and LAN management is used in 10 Base-T, 100 Base-T and Gigabit Ethernet systems.

Cheapernet

Cheapernet also follows IEEE 802.3 but provides a low-cost alternative to Ethernet in which the transceiver function is incorporated within the terminal equipment (the cable tap box and transceiver cable are thus no longer required). Savings in cost are also made by using lower grade coaxial cable throughout the network and terminal equipment is simply attached using T-connectors at strategic points. Cheapernet is thus very much simpler to install than its more expensive counterpart.

The limitations of Cheapernet are that its segment length is restricted to 185 metres (making repeaters essential for larger networks) and that the IEEE 802.3 cable/terminal ground isolation scheme (which requires d.c. isolation between transceiver and terminal) is not so easy to implement using VLSI devices when the transceiver function is to be integrated within the controller. These two drawbacks can seriously limit the cost-effectiveness of the system when compared with a full Ethernet implementation of IEEE 802.3.

Baseband IBM PC LAN

The baseband IBM PC LAN allows a mixture of PCs and PS/2 machines to communicate with one another at relatively low-cost. A PC network adaptor/A, network support program, and PC LAN program is required at each node. CSMA/CD protocol is used and the network employs a twisted pair cable in which the data rate is 2 Mbps.

Broadband IBM PC LAN

The broadband IBM PC LAN also allows a mixture of PCs and PS/2 machines to communicate with one another with the aid of a PC network adaptor II/A, network support program and PC LAN program at each node. In addition, one or more PC network translator units are required. Each translator unit can handle up to eight nodes at distances not exceeding 200 feet. Larger networks can be realised using further translators and standard IBM Cable System components. The broadband PC LAN operates at a data rate of 2 Mbps with CSMA/CD protocol and coaxial cable.

IBM Token Ring LAN

The IBM Token Ring LAN can be used to implement a lager network in which computer systems (eg, System/370) can communicate with a variety of PC and PS/2 machines. Token ring network adaptors are required at each PC or PS/2 node together with one or more multi-station access units. The data transfer rate for the network is 4 Mbps.

ICL Macrolan

ICL Macrolan employs a modified form of token-passing ring which incorporates features designed to improve the efficiency and fault tolerance of the system. The system uses optical fibres as the physical medium with multi-port ring switches which permit disconnection of inactive nodes. Two optical cables are required for each node in order to permit full duplex operation.

Manufacturing Automation Protocol (MAP)

Manufacturing Automation Protocol (MAP) is a broadband token bus system which uses community television (CATV) coaxial cable as its physical medium. The standard was developed by General Motors but has become widely accepted by a number of major manufacturing and production engineering concerns as a robust and versatile factory networking standard.

The layers within MAP closely follow the ISO model for OSI (there is direct correspondence at the application, session, transport, network, data link and physical levels). The data link layer follows IEEE 802.2 while the physical layer corresponds to the IEEE 802.4 (token bus) standard. The system employs quadrature amplitude modulation (QAM) at a data rate of 10 Mbps.

Technical and Office Protocol (TOP)

Technical and Office Protocol (TOP) was developed by Boeing Computer Services and has much in common with MAP. TOP can, however, be implemented at lower cost using the CSMA/CD protocol defined under IEEE 802.3. The upper layers of TOP correspond closely to those within MAP and thus it is possible to interwork the two systems. TOP version 1.1 employs standard Ethernet trunk coaxial cable with a maximum segment length of 500 metres (adequate for most office and commercial environments). A routing device (or *router*) permits interconnection of MAP and TOP networks. The routing device essentially provides a bridge above layer of the ISO model and resolves any differences between address domains, frame sizes, etc.

Summary of popular LAN specifications

Name	Supplier	LAN type	Topology	Transmission medium	Protocol	Maximum cable length	Maximum nodes	Data rate (bps)
Apple Talk	Apple Computer	Ba	Bus	T/pair	CSMA/CD	300 m	32	230
Cambridge	Camtec	Ba	Bus	T/pair F/opt	Cambridge Ring			10M
Ethernet	various	Ba	Bus	Coax	CSMA/CD	500 m	100	10M
FastLAN	Wang	Br	Bus	Coax	CSMA/CD	186 m		10M
IBM PC Network	IBM	Br	Bus	Coax	CSMA/CD	300 m		2M
Isolan	BICC	Ba	Bus	Coax	CSMA/CD	4 km	1024	10M
Netware	Novell	Ba	Bus/star	Coax	CSMA/CD		100	10M
NIM 1000	Olivetti	Ba	Bus	Coax	CSMA/CD	2.5 km	1024	10M
Nimbus Network	Research Machines	Ba	Bus	Coax	CSMA/CD	1.2 km	32+	800
OSLAN	ICL	Ba	Bus	Coax	CSMA/CD			10M
Planet	Racal-Milgo	Ba	Ring	Coax	Token	300 m	500	10M
Primenet	Prime Computer	Ba	Ring	Coax	Cambridge Ring	2.25 km	63	10M
WangNet	Wang	Br	Bus	Coax	CSMA/CD	13 km		10M
X.25-based	various	Ba	Star	T/pair F/opt	X.25			64 k

Notes:

Ba = baseband   T/pair = twisted pair
Br = broadband   F/opt = fibre optic

Twisted-pair Ethernet

10base-T

This carries 10 Mbps baseband data over twisted pair. The maximum segment length is 100 m, with a maximum of 1024 nodes. The velocity constant for twisted pair cable ranges from 0.59 to 0.63. For delay calculations, a transmission velocity of 200 m/µs may be assumed. The characteristic impedance is about 100 ohm.

Manchester encoded data is transmitted, which is one of the simplest and allows the receiver to recover the clock synchronisation. One problem with baseband transmission is that the cable attenuates the signal and can make it difficult to separate data from the noise. This problem does not arise over the relatively short cable distances used on most LANs.

Data is transmitted as an Ethernet packet comprising a certain number of bytes (each byte is 8-bits long). The maximum data payload is 1500 bytes and longer files or packets must be broken down into smaller segments before transmission.

Frame structure

Bytes	Field
7	Preamble
1	Start of frame delimiter
2 or 6	Destination Address
2 or 6	Source Address
2	Length of data field
0 to 1500	Data
0 to 46	Pad
4	Checksum

There is a minimum packet size. Valid frames must be at least 64 bytes from the start of the destination address to checksum. If the data portion of the frame is less than 46 bytes, padding is used to bring the frame up to the minimum value. The limitation on the maximum frame length (12.2 kbits) is to ensure fair access to all users.

100base-T

The drivers for the introduction of 100 Mbps LANs were: increased Internet usage, increased file sizes, video conferencing, and the increase in e-mail usage. These applications, coupled with higher processing speeds and broadband transmission systems (such as FDDI, SONET, ATM and ADSL) made faster LANs essential. The IEEE decided to supplement the existing 802.3 standard by increasing the data rate and maintain current protocols. The standard was designated as 802.3U and has the name 100base-T; meaning 100 Mbps, Baseband, Twisted-pair copper cabling.

Implementation of 100base-T uses either UTP or STP copper cabling, or optical fibre. There are no standards for using co-axial cable. The older (pre 1988) Category 3 (AWG 24) twisted-pair copper cabled structures do not use Manchester coding, instead they use Ternary (3-level) coding, which is designated 8B6T. Eight bits are translated into six ternary symbols. Using Category 5 (AWG 22/24) twisted-pair copper cabling do not use Manchester coding either, instead they use binary coding, which is designated 4B5B (4 bits are translated into 5 binary symbols, which means that the cable data rate exceeds the system data rate). In this scheme, 100baseTX can clock data over the cable at up to 125 Mbps and can perform as full duplex at 100 Mbps (transmit and receive).

LANs carrying a mix of 10 Mbps and 100 Mbps are possible by using switching via high-speed back-planes. Thus is it possible to build a network with additional workstations having 100 Mbps line cards, whilst existing terminals continue to operate at 10 Mbps.

Gigabit Ethernet

Gigabit Ethernet uses the same frame format and support for CSMA/CD (carrier sense multiple access with collision detection) protocol, full duplex, flow control, and management objects as defined by the IEEE-802.3 Ethernet and Fast Ethernet standard.

Gigabit Ethernet 1000base nomenclature:

1000base-SX	850 nm multimode fiber
1000base-LX	1300 nm multimode and single-mode fiber
1000base-CX	Short-haul copper ('twin-axial' STP)
1000base-T	Long-haul copper over UTP

The following table shows the cable types and maximum lengths that are supported by the standard:

Ethernet cable types

	Ethernet 10base-T	Fast Ethernet 100base-T	Gigabit Ethernet
Data Rate	10 Mbps	100 Mbps	1 Gbps
Category 5 UTP	100 m (min)	100 m	25–100 m
STP/Coax	500 m	100 m	25 m
Multi-mode fibre	2 km	412 m (hd)/2 km (fd)	500 m
Single-mode fibre	25 km	20 km	2 km

Note: hd = half-duplex; fd = full-duplex

Ethernet (co-axial cable)

Basic Ethernet connecting arrangement

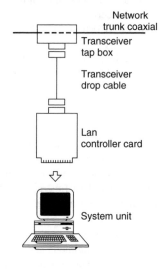

Ethernet transceiver cable pin connections

Pin number	Assignment
1	shield (also connected to connector shell)
2	collision+
3	transmit+
4	reserved
5	receive+
6	power
7	reserved
8	reserved
9	collision
10	transmit
11	reserved
12	receive
13	power+
14	reserved
15	reserved

Ethernet transceiver cable specifications

Construction:	four-pair 78 ohm differential impedance plus overall shield (eg, BICC H9600)
Loop resistance:	less than 4 ohm for power pair
Signal loss:	less than 3 dB at 10 MHz (typical maximum length equivalent 40 metres)
Connectors:	1 × female and 1 × male 15-pin D-connector

Typical Ethernet interface configuration

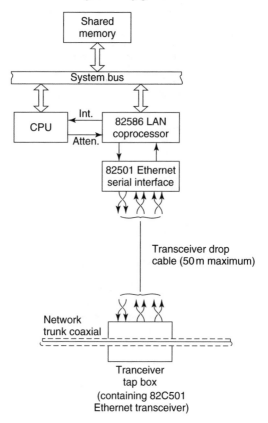

Typical Cheapernet interface configuration

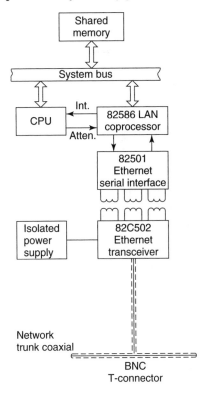

Internal architecture of the 82C502 Ethernet transceiver

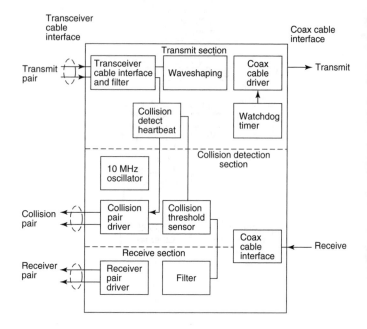

Wireless LANs

To allow simpler provision and movement of LAN based equipment, wireless connections are becoming popular. Wireless LANs (IEEE 802.11), Bluetooth and the higher speed HIPERLAN are all described here. The use of infrared light, instead of radio signals is also discussed.

Bluetooth, Wireless LAN and HIPERLAN nodes communicate through radio based devices, which are mostly plug-in cards (ISA, PCMCIA) fitted into personal computers. The radio device has two main functions: the radio modem and the media access controller (MAC).

The radio modem modulates the data onto the radio frequency carrier and thus transmits radio signals. It also receives and demodulates transmissions from other modems. It is composed of antenna(s), amplifiers, frequency synthesisers and filters. The carrier frequency, bandwidth and transmit power are all controlled by the appropriate standard.

To overcome noise and to increase the reliability of wireless LAN systems, diversity in terms of frequency, space or time is used. Spread spectrum is a form of frequency diversity because it uses more bandwidth than necessary to avoid noisy parts of the spectrum. Retransmission and Forward Error Correction (FEC) of the transmitted signals give temporal diversity. Spatial diversity in a radio system is achieved by using two or more antennas, to provide two paths for the radio signal.

Media access control (MAC)

The MAC is responsible for running the signalling protocol, which is also determined by the appropriate standard. The main characteristics of the MAC protocol are packet format (size, headers), channel access mechanisms and network management. The two channel access mechanisms used by the MAC protocol in wireless LAN systems are carrier sense multiple access/collision avoidance (CSMA/CA) and Polling MAC.

CSMA/CA is the channel access mechanism used by most wireless LANs in the ISM bands. A channel access mechanism is the part of the protocol that specifies when to listen, when to transmit. The basic principles of CSMA/CA are listen before talk and contention. This is an asynchronous message passing mechanism (connectionless), delivering a best effort service, but no bandwidth and latency guarantee. Its main advantages are that it is suited for network protocols such as TCP/IP, adapts quite well to variable traffic conditions and is quite robust against interference.

CSMA/CA is derived from CSMA/CD (collision detection), which at the heart of Ethernet MAC: the main difference between them is the carrier sense function. On a wire, the transceiver has the ability to listen whilst transmitting and hence can detect collisions. But a wireless system cannot listen on the channel whilst transmitting, since transmit and receive frequencies are the same and because the transmit level is far higher than the receive level. Therefore the wireless MAC protocol tries to avoid, instead of detecting, collisions.

The protocol starts by listening on the channel (this is called carrier sense) and, if the channel is found to be idle, sends the first packet in the transmit queue. If it is busy, due to either another node transmission or interference, the node waits until the end of the current transmission and then starts the contention (waits a random amount of time). When its contention timer expires, if the channel is still idle, the node sends the packet. The node having chosen the shortest contention delay wins and transmits its packet. Because the contention is a random number, each node is given an equal chance to access the channel (on average).

A form of carrier sense used by some systems is request to send/clear to send (RTS/CTS). The RTS/CTS is a handshaking protocol: before sending a packet, the transmitter sends a RTS and waits for CTS from the receiver. The reception of CTS indicates that the receiver is able to receive the RTS, so the packet may then be transmitted (the channel is clear in its area). Any node within range of the receiver hears the CTS and so knows that a transmission is about to take place.

The RTS and CTS messages contain the size of the expected transmission, so any node listening will know how long the transmission will last. This is useful because the data transmission itself may not be heard. The use of RTS/CTS lowers the overhead of a collision on the medium, because RTS collisions are much shorter in time. If two nodes attempt to transmit in the same slot of the contention window, their RTS collide and they have to try again. They loose an RTS instead of a whole packet.

Polling is a major channel access mechanism. The 802.11 standard offers a polling channel access mechanism (point co-ordination function) in addition to the CSMA/CA one. Polling is a mixture of both time division multiple access (TDMA) and CSMA/CA. TDMA is not used because although it is ideal for voice traffic (which is why it is used by GSM) it is not suitable for IP traffic that occurs in bursts. In the polling access scheme, the base station retains total control over the channel but the frame content is no longer fixed, allowing variable size packets to be sent. The base station sends a short 'poll packet' to trigger the transmission by the node. The node just waits to receive a poll packet and, upon reception, starts data transmission.

Error control

There is a higher error rate on the radio link than over a wire and this leads to packets being corrupted. Packet losses at the MAC layer cause problems for TCP, so most MAC protocols implement positive acknowledgments and MAC level retransmissions. Each time a node successfully receives a packet, it immediately sends back a short message (an ACK) to the transmitter. If the transmitter does not receive an ACK within a certain period after sending a packet, it will retransmit the message.

Some wireless LAN systems use fragmentation to increase the data throughput on an error-prone radio channel. Fragmentation involves breaking down big packets into small pieces before transmitting them. This adds some overhead, because packet headers are duplicated in every fragment. Each fragment is individually checked and retransmitted if necessary. In case of a corrupted packet being received, the node need only retransmit one small fragment, so it is faster.

Industrial scientific and medical (ISM) band

Wireless LANs and bluetooth both use the industrial scientific and medical (ISM) frequency band at 2.4 GHz. The ITU has decreed that this band may be used for ISM purposes in all parts of the world. However, national regulators deal with specific licensing. In Europe, all equipment operating in these bands must comply with ETSI standard ETS 300 328. In the USA, any WLAN operating in this band must comply with FCC part 15. In Japan, compliance with MPT (Ministry of Post & Telecommunications) ordinance 79 is necessary. However, ISM bands are unlicensed, which means that a large number of other users may be using the same frequencies. The 2.4 GHz band also suffers from microwave oven radiation.

The ISM band regulations specify that spread spectrum techniques have to be used (either direct sequence or frequency hopping). These techniques spread the signal over a large bandwidth to reduce localised interference. The radio modem used for direct sequence is more complicated than the frequency hopping one, but the direct sequence method requires a simpler media access control (MAC) protocol. Frequency hopping is more resistant to interference, but direct sequence offers better performance when multi-path propagation is a problem. Frequency hopping is normally used.

The ISM band regulations limit the radio bandwidth to 1 MHz for frequency hopping systems. The available data rate can be increased by complex modulation schemes, allowing several data bits per symbol. This means that the receiver has to distinguish between a number of different symbols. To do this, the signal-to-noise ratio of the received signal has to be higher than if a simple two-symbol system were operating. Since the various standards limit the transmitter power level, the operating range is reduced.

Bluetooth

Bluetooth also operates in the 2.4 GHz license-exempt Industrial, Scientific and Medical (ISM) band. Bluetooth uses frequency hopping and hops between 79 carriers spaced 1 MHz apart. Pseudo-random hop sequences are used so that each carrier frequency is used with equal probability. Gaussian minimum shift key (GMSK) modulation is used on these carriers. Compared to IEEE802.11 wireless LANs, Bluetooth uses a very fast hop rate; 1,600 hops per second. This means it stays on each frequency for a 625 μs time interval, which is known as a slot.

The bluetooth protocol is a combination of circuit and packet switching. Slots can be reserved for synchronous packets and each packet is transmitted in a different hop frequency. The duration of a packet nominally covers a single slot, but can be extended to cover up to five slots. Bluetooth uses TDD (time division duplexing), which means that transmit and receive packets are carried in alternate slots. Bluetooth can support either an asynchronous data channel and up to three simultaneous synchronous voice channels, or a single channel which simultaneously supports asynchronous data and synchronous voice.

The bluetooth packet format allows one packet to be transmitted in a slot. Each packet consists of an access code, a header and data payload. The access code is 72 bits long, the header is 54 bits long and the payload is of variable length; between 0 and 2745 bits long. Slots can be combined; a packet can be one, three, or five slots in length. Multi-slot packets are transmitted on the same frequency carrier, before the transmitter continues with the hop sequence. This reduces the transmission time lost in changing frequencies and reduces the control overhead (a five slot packet has only one access code and header, where before there were five).

Synchronous connection oriented (SCO) links support symmetrical, circuit-switched, point-to-point connections typically used for voice. These links are defined on the channel by reserving two consecutive slots (forward and return slots) at fixed intervals. The fixed interval size depends on the level of error correction required. Three kinds of single-slot voice packets have been defined, each of which carries voice data at 64 kbit/s. Voice is usually sent unprotected, since the CVSD voice-encoding scheme is very resistant to bit errors. If the interval is decreased, FEC rates of 1/3 or 2/3 can be selected.

Asynchronous connection-less (ACL) links support symmetrical or asymmetrical, packet-switched, point-to-multi-point connections typically used for burst data transmission. 1-slot, 3-slot and 5-slot data packets are defined. Data can be sent either unprotected or protected by a 2/3 FEC rate. The maximum data rates are obtained when an unprotected 5-slot packet is used.

Type of packet	Symmetric (kbit/s)	Asymmetric (kbit/s)
1-slot (protected)	108.8	108.8/108.8
1-slot (unprotected)	172.8	172.8/172.8
3-slot (protected)	256.0	384.0/54.4
3-slot (unprotected)	384.0	576.0/86.4
5-slot (protected)	286.7	477.8/36.3
5-slot (unprotected)	432.6	721.0 (max)/57.6

The packet definitions have been kept flexible as to whether or not to use FEC in the payload. The packet header is always protected by a 1/3 rate FEC, this is because it contains valuable link information that needs to survive bit errors. For data transmission, an ARQ scheme is applied.

The 5 GHz band (HIPERLAN and IEEE802.11)

HIPERLAN and satellite systems use the 5 GHz band. The band from 5.15 to 5.25 GHz (three radio channels) is available across Europe, with 5.25 to 5.35 GHz (two extra channels) also available in some countries, but not in the UK. These bands may be used *indoors* by both HIPERLAN/1 and HIPERLAN/2, and transmitted power is limited to 200 mW.

HIPERLAN/2 systems use the band 5470–5725 MHz, both indoors and outdoors, although transmitted power is limited to 1 W. In order to co-exist with satellite feeder links, HIPERLAN/2 systems using this band must incorporate power control and dynamic frequency selection. HIPERLAN/1 systems do not have these facilities and therefore cannot be used here since they would risk causing interference to satellite systems.

In the USA, three U-NII bands are specified. These have very liberal rules – spread spectrum is not mandated. No channels have been allocated and there are different power maximums, depending on the band being used. The low band covers 5.15 to 5.25 GHz, the mid band covers 5.25 to 5.35 GHz and the high band covers 5.725 to 5.825 GHz.

In Japan, 5.725–5.875 GHz is set aside for ISM applications, such as wireless LANs.

In the 5 GHz band, higher speeds are possible because of the availability of more bandwidth. This is typically 10 to 40 Mb/s (which in theory is also available in the 2.4 GHz band). The disadvantage with

PHY modes of 802.11 and HIPERLAN/2

Modulation code	Rate	Net rate
BPSK	$1/2$	6 Mbps
BPSK	$3/4$	9 Mbps
QPSK	$1/2$	12 Mbps
QPSK	$3/4$	18 Mbps
16-QAM	$3/4$	36 Mbps
HIPERLAN/2 only		
64-QAM	$2/3$	48 Mbps
IEEE 802.11 only		
64-QAM	$3/4$	54 Mbps

using higher frequencies is a reduced range and increased sensitivity to obstacles.

Both the IEEE and ETSI standardisation bodies have worked together in order to harmonise the physical layer for 5 GHz. The PHY layer offers the transmitting and receiving service on the wireless medium. It uses orthogonal frequency division multiplexing (OFDM) with 48 active sub-carrier plus 4 sub-carrier for pilot symbols using an FFT size of 64. The operating frequency is between 5 and 6 GHz with a bandwidth of 20 MHz per frequency channel.

OFDM does not use a single carrier nor employ frequency hopping nor use a spreading code. Instead, it simultaneously uses a large number of narrow carriers (e.g. 48) in a radio channel (20 MHz). The data is divided into several interleaved, parallel bit-streams, and each one of these bit streams modulates a separate sub-carrier. Each sub-carrier can be modulated using BPSK, QPSK, or QAM. These sub-carriers all are used for one transmission link between a mobile and an access point. One of the benefits of OFDM is the robustness against the adverse effects of multi-path propagation, common in cluttered indoor environments.

Infrared

The IEEE 802.11 provides for an infrared (IR) physical layer. Instead of a radio channel, this uses infrared light at a wavelength of 850–950 nm. The light source is an LED, which is safety rated as Class 1 (eye safe). Data modulates the LED using pulse position modulation (PPM) and achieves data transmission rates of 1 or 2 Mbps.

Infra-red is intended for indoor environments, with a typical range of 10–20 m (in favourable conditions). The light is not a focused beam, but instead is diffuse with reflections off walls and ceilings, so that line-of-sight is not required. However, a cell is limited to a single room because IR will not penetrate walls and is attenuated by glass.

Fibre distributed data interchange (FDDI)

FDDI is typically used on university campus or business premises and gives a 100 Mbps wide backbone for LANs. Up to 500 nodes can be supported, spaced no more than 2 km apart. Two fibre optic rings are used and these are arranged so that data is counter rotating – the same data travels on both of the fibre rings, one clockwise and the other anti-clockwise. This topology gives some fault tolerance – the dual ring is converted to single ring if a fibre fails, see the diagram below. The maximum ring length is 100 km as a dual ring (200 km as a single ring).

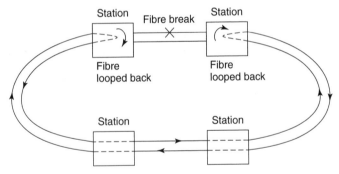

Dual Ring Fault Tolerance

ANSI is the main standards authority for FDDI and they have given it the designation X3T12. The international standards organisation have also 'standardised' it under the designation ISO 9314. The FDDI standard was developed from the token ring standard, IEEE-802.5. In the ANSI standard there are four key components: MAC – media access control; PHY – physical layer; PMD – physical media dependent; and SMT – station management protocol. The MAC component defines addressing, scheduling, data routing and communication using protocols such as TCP/IP. The PHY component handles encoding/decoding, NRZI modulation and clock synchronisation. The PMD component handles analogue base-band transmission between nodes–fibre and copper. The SMT component handles ring management including neighbour identification, fault detection and reconfiguration.

FDDI cables usually employ multi-mode fibre with a 62.5 micron core and 125 micron cladding diameter. The 62.5/125 fibre is favoured because low-cost LED/photodiode technology can be used to drive/detect light in wider fibre. However, 50/125, 85/125 and single mode fibre can also be supported. Four fibres are used in each cable (2 transmit and 2 receive) although the installation of spares is recommended for replacement of faulty fibres.

It is also possible to use FDDI formatted data over copper, which has the acronym CuDDI. This uses copper unshielded twisted pair (UTP) or shielded twisted pair (STP) cables for connecting to local terminals. The copper cable can be up to 100 metres in length and ANSI standard TP–PMD (twisted pair–physical medium dependent) applies. The advantages of using twisted pair cable are that it is low cost and that installation and termination are simpler. Also, copper based transceivers are cheaper, smaller and require less power compared to fibre-based systems.

Unlike token rings defined by IEEE 802.5, there is no active ring monitor. Instead, each ring interface has its own clock synchronised to incoming data. The outgoing data is transmitted using a local clock. FDDI is not synchronous but is plesiochronous.

All data is encoded prior to transmission and this uses a 4-out-of-5-group code known as 4B/5B. In this scheme, every 4-bit group (16 different combinations) is mapped onto a 5-bit code (symbol). The 5-bit symbols for 4-bit data groups are chosen such that no more than two successive zeros occur. Certain 5-bit symbols that are not used for data encoding are used instead as control symbols. The 5-bit symbols are passed through a NRZI (non return to zero inverter) which produces a signal transition for logic 1 and no change for logic 0. The 4B/5B encoding combined with NZRI modulation guarantees that there is one signal transition at least for every three bits transmitted.

The control symbols are abbreviated to characters I, H, Q, J, K, T, R, S and L. Control symbols I, H and Q give the fibre state: I = idle = 11111, H = halt = 00100, Q = quiet = 00000. Symbols J, K and T are used as frame delimiters. Symbols R and S are logical indicators ('0', '1'), and L is reserved for FDDI version II.

In an idle FDDI system, a 'token' is transmitted around the ring continuously. Each station receives the token and then re-transmits it into the ring. When a station wishes to transmit a data packet, it must 'capture' the token first. A token is captured when it has been received by a station, but not re-transmitted. Once a station holds the token it has permission to transmit data packets. After the data packets are transmitted into the FDDI ring, the station 'releases' the token by transmitting it into the ring. Further transmissions are not possible until the token is captured again.

The token format comprises a preamble, start delimiter, frame control and an end delimiter. The Pre-Amble (PA) contains 16 or more Idle (I) symbols, which produce line changes at the maximum frequency. The start delimiter (SD) contains J and K symbols to enable the receiver to fix the correct symbol boundaries. The frame control (FC) contains 2 symbols that determine the type of information carried in the data frame and the end delimiter (ED) contains 2 T symbols.

The data frame (data packet) is shown below. The header comprising PA and SD symbols, which are exactly the same as for a token. The FC symbols describe the frame type and features such as whether it is synchronous, asynchronous and the address field size. The data may contain MAC, SMT or LLC information depending on a symbol set in FC. Address information is given in the destination and source address (DA and SA), which is held in 4 or 8 symbols (as set by FC).

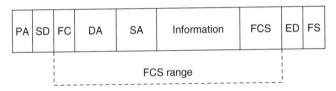

FDDI data frame

The data frame information field can be empty or contain an even number of symbols, up to a maximum of 9000 symbols (or 4500 bytes) including all the fields.

The data frame ends with frame check sequence (FCS) which is 8 symbols long, an end delimiter (ED), which is 1 symbol and frame status (FS), which checks the frame validity and reception. The following indicators are defined in the FS field: E = error detected, A = address recognised, and C = frame copied. Other symbols may be added possibly followed by a T symbol.

FDDI II is an extension of FDDI to support isochronous traffic. Isochronous service is required when there are strict timing constraints, such as multimedia traffic where data, digitised sound, graphics and video are integrated. Two modes of operation are supported: basic, which is FDDI, and hybrid, which is FDDI plus isochronous.

Information is carried in periodic frames called cycles and one cycle is generated every 125 microseconds. At 100 Mbps a 125 microsecond cycle can carry 12000 bits or 3125 symbols. A cycle is divided into five preamble symbols, followed by a 24 symbol (12 byte) cycle header, 24 symbols for packet type data channel dedicated packet group (DPG) and 16 wide-band channels (WBCs), each 96 bytes wide or 3072 symbols long. In total there are 3125 symbols.

The cycle header format is illustrated below. It includes 2 symbols for the start delimiter (SD), 1 symbol each for the synchronisation control (C1) and sequence control (C2), 2 symbols for the cycle sequence (CS), 16 symbols (P0 to P15) containing the WBC programming template and 2 symbols for the isochronous maintenance channel (IMC). The L symbol (FDDI II only) is used to ensure the uniqueness of the cycle delimiter, SD, within the cycle packet type. I and L, instead of J and K delimit data.

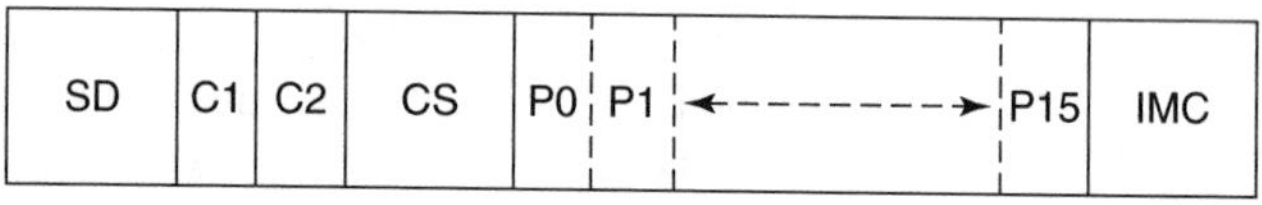

FDDI II cycle header

9 Wide area networks

Wide area networks (WANs) occur when two or more local area networks (LANs) are joined over a long-distance link. This link requires a digital access from the local telephone exchange, and this can be provided by a conventional (e.g. V.34) modem, a DSL modem or ISDN terminal. Where an E1 or T1 trunk is provided, data is often transmitted in ATM packets.

Connecting LANs

The connection between a LAN and an external transmission path is via a gateway. The gateway is either a router or a server fitted with line termination cards. When it is necessary to transmit data over the 'wide-area' link, the gateway opens a connection and performs all the necessary data rate and protocol adaption. Usually, the wide-area link is not capable of transmitting data as fast as it is transmitted over the LAN. For example, a 100 Mbps LAN may only have a 2 Mbps wide-area link. Data is collected by the gateway at the LAN speed and then forwarded at the speed of the wide-area link. Gateways are also used to provide a 'firewall', to control access and ensure that viruses are not transferred between LANs.

Integrated services digital network

Most telecommunications providers use an integrated digital network. However, the service extended to the user is often analogue, with digital to analogue conversion taking place in the local exchange. Digital service is available, but the cost is usually higher because expensive line termination equipment is required. The digital service is called integrated services digital network (ISDN). The International Telecommunications Union – Telecommunications sector (ITU-T) has published a series of recommendations for ISDN, called the I-series. The I.100 series gives general information, such as basic descriptions and definitions of terms. The I.200 series describes services, whilst the I.300 series describes the network capabilities.

There are two systems: basic rate and primary rate ISDN. Basic rate provides two 64 kbps circuits, known as B-channels (B = basic) and is known as ISDN2. These circuits can be coupled to provide a single 128 kbps circuit. A 16 kbps control (or signalling) channel is also provided and this is known as the D-channel. Thus ISDN2 is also referred to as 2B + D, basic rate ISDN or even IDSL.

Primary rate ISDN is defined under ITU-T recommendation I.431 as a 30-channel service in Europe or a 23-channel service in the USA and Japan. Transmission of the ISDN signals between the customer and the local exchange is usually over either optical fibre or copper pairs using HDSL; rarely, a coaxial cable is used to connect the customer premises with the local exchange.

The European version of primary rate ISDN is also known as ISDN30. This provides thirty 64 kbps B-channels and a 64 kbps D-channel. The bearer for this service is a 2.048 Mbps E1 link. Timeslot 0 (TS0) is used for synchronisation. TS1–TS15 are used for ISDN B-channels numbered 1 to 15. TS16 is used for the signalling channel (the D-channel). TS17–TS31 are used for ISDN B-channels 16 to 30.

The US version of primary rate ISDN provides twenty three 64 kbps B-channels and one 64 kbps D-channel. The bearer for this service is a 1.544 Mbps T1 link.

High-speed digital subscriber line (HDSL)

The HDSL system was developed to provide a means of extending 2 Mbit/s E1 or 1.544 Mbit/s T1 data to subscribers when the only access cable is twisted pair. A typical application is the provision of multiple telephone circuits to private branch exchanges (PBXs) or high speed local area network (LAN) connections. Current systems require two or three pairs; the data is shared between the pairs so that the data rate on each pair is reduced, and this allows the line length to be up to 3 miles (5 km).

The data on each pair is encoded by the 2B1Q method. Two binary bits (2B), or dibits, are converted into one of four voltage levels (1Q). The DC voltage levels are transmitted along each pair and measured at the receiver for decoding back into dibits. This is fairly straightforward, except that data is simultaneously being transmitted in the opposite direction. Echo cancellation techniques are used to subtract the transmitted signal from the composite signal across the line, thus leaving the received signal. A sample of 2B1Q data is shown in the diagram below.

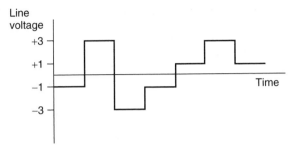

2B1Q data

Asymmetric digital subscriber line (ADSL)

The ADSL system was developed as a result of competition from cable television companies. Cable TV companies provide telephone services alongside their television service, often at low-cost, as a marketing tool to sell their product. The ADSL system allows incumbent telephone operators to use existing twisted pair cables to provide television services; the aim is to reduce the number of customers wishing to change service provider. The system is asymmetric because the data rate from the exchange to the subscriber is much greater than in the opposite direction. A data rate of 2 Mbit/s is required in order to transmit compressed video signals. Such an asymmetric system would also be suitable for Internet access, where far more data flows from the ISP than from the Internet subscriber.

The ADSL system does not use 2B1Q line coding. The design criteria for ADSL is that is should use the existing telephone service pair, thus the telephone must operate as before using a DC loop to indicate an active line. Since the 2B1Q line coding uses DC voltage levels, it is incompatible with DC loop conditions. The ADSL system uses the spectrum above audio frequencies for transmission: the ADSL system is a wideband modem.

There are two modulation methods. One is called discrete multitone (DMT) and uses a number of discrete carrier frequencies, each one being modulated by part of the data. The modulation spectrum of each carrier occupies a 4 kHz bandwidth. This is like having lots of modems running in parallel, each with a different carrier frequency. The second modulation method is called carrier-less amplitude and phase (CAP) and is similar to quadrature amplitude modulation (QAM). Instead of having a carrier that is amplitude and phase modulated, the CAP system synthesizes an equivalent waveform.

The disadvantage of using high frequencies for transmission is that line attenuation increases with frequency. This means that those signals which use the higher frequencies are attenuated more than those using frequencies just above audio and the working range is limited. The advantage that ADSL has over HDSL is that only a single pair is required. The disadvantage is that because of line attenuation considerations, high level signals are transmitted in the high data rate direction. To prevent near-end crosstalk problems, the high data rate direction must always be from the exchange to the customer.

ADSL (G.lite)

A low-cost version of ADSL avoids the need for a filter in the customer premises to separate the POTS and ADSL signals. It is known as splitter-less ADSL and is defined by ITU recommendation G.992.2. The line modulation uses DMT and allows a data rate up to 1.5 Mbps in the downstream direction towards the subscriber and up to 512 kbps in the upstream direction towards the service provider.

Conventional ADSL using DMT modulation occupies 256 frequency bands, each 4 kHz wide. These bands start at 25 kHz and extend up to 1.049 MHz. The alternative G.lite uses 96 frequency bands and therefore its line borne signal has less high frequency content than standard ADSL, with a maximum frequency of 409 kHz.

Single-pair digital subscriber line (SDSL)

Due to the lack of a formal naming convention in the telecommunications industry, the term SDSL has become more generic over time and to some means any symmetric DSL system. In this section, SDSL is given to be a symmetrical digital subscriber line system developed under the auspices of ETSI that works over a single pair, at variable data rates up to 2.3 Mbps. It is intended to carry E1 traffic.

The line modulation is 2B1Q and this is commonly used with Maximum likelihood estimation at the receiver. The 2B1Q signalling system transmits one symbol for every two data bits and this is used by ISDN (sometimes called ISDL, since it is an xDSL technology). However, at the higher data rates employed by SDSL, the pulses are distorted and determining which symbol was transmitted becomes more difficult. Maximum likelihood (ML) estimation techniques are used to determine the most likely symbol. Pulse distortion is worse on long subscriber lines, so the maximum line length for SDSL is about 3 km. At T1 data rates, the maximum line length is about 3.4 km, which is why ANSI started to develop their own system called HDSL2.

ETSI have produced a modified SDSL standard that uses the same 16 level PAM line codes as used by the ANSI HDSL2 standard, and which has been incorporated into the ITU G.991.2 standard. ETSI SDSL systems using 2B1Q line codes will eventually be replaced by SHDSL (G.shdsl or G.991.2) systems.

High-speed digital subscriber line 2 (HDSL2)

Operation of HDSL over a single pair is possible using HDSL2. This is intended to carry fixed rate T1 (1.544 Mbps), like HDSL. A 16-level PAM line code is used, with spectral shaping. The symbol rate transmitted over the line is 517.3 k symbols per second. Each symbol represents four bits: three data bits and one forward error correction (FEC) bit. The FEC process uses a single-stage 512-state Trellis-coded modulation. The power spectral density of the modulated signal is not flat, but uses OPTIS shaping.

HDSL2 is a single-pair, ANSI standard-based replacement for HDSL, which is really only applicable in North America. ANSI HDSL2 will be replaced by SHDSL (G.shdsl or G.991.2).

Single-pair HDSL (G.shdsl–G.991.2)

Standardised by the ITU as G.991.2, SHDSL is intended to provide symmetrical data transmission at rates of up to 2.3 Mbps. Variable data rates will be possible, from 144 kbps to 2.3 Mbps. Like the ANSI HDSL2 system and the revised ETSI SDSL specification, a 16-level PAM line code is used. However, the G.hsdsl system does not apply spectral shaping to the signal before it is transmitted: the power spectral density is flat. This allows the maximum line length to be 30% greater than 2B1Q SDSL and 20% greater than for HDSL2.

Very-high-speed digital subscriber line (VDSL)

Very high speed digital subscriber line (VDSL) systems are not yet standardised. The purpose of VDSL is to provide a very high data rate into offices and buildings. The system will be asymmetric, in the early models at least, allowing a maximum upstream rate from the customer to the exchange of about 2 Mbit/s. The downstream data rate is expected to be up to 55.2 Mbit/s, which limits the range over twisted pair cables to about 300 m. Lower data rate systems permit a longer range over twisted pair cable. Limiting the downstream data rate to 27.6 Mbit/s will allow a range of 1 km. Further limiting the rate to 13.8 Mbit/s allows a working range of 1.5 km.

Transmission of data to outside the building will use optical fibre as a 'fibre to the kerb' (FTTK) system. VDSL will perform the final drop into the customer's premises.

Asynchronous transfer mode

Asynchronous transfer mode

Major standards organisations (such as ISO, ITU and ANSI) have recognised ATM as the preferred standard for Broadband Integrated Services Digital Networks (BISDN). Like X.25 and Frame Relay, Asynchronous Transfer Mode (ATM) is a packet-switching technology. ATM offers considerable promise for the future of broadband digital data services. The real benefits of ATM will be realised whenever video, digital data, and voice services are all integrated on the same wide area network (ie, the much heralded 'information superhighway').

ATM cells

ATM packet cells comprise a total of 53 bytes of which 48 are data (the payload) and 5 are overhead (the header). Depending upon the type of data being transmitted, the 48 bytes of data may contain additional overhead required to partition and reassemble longer messages. Headers contain Virtual Channel Identifiers (VCI) and Virtual Path Identifiers (VPI). These are modified, as required, by each node in the network in order to ensure that the cell reaches its destination.

ATM is an asynchronous protocol in that packets can be received at arbitrary intervals of time. Despite this, the ATM bitstream is both continuous and synchronous. This may at first appear to be something of an anomaly but you simply need to recall that packets can appear randomly distributed in time and when packets are not present, other bits are transmitted to represent the idle state. Thus it is the packet cells that are asynchronous not the bitstream itself.

ATM can operate at any speed and most high-speed network standards can support ATM. Data rates of 51.84 Mbps (OC-1 and STS-1) or 155.52 Mbps (OC-3 and STS-3) are commonly used with the ANSI Synchronous Optical Network standard (SONET). Note that 'OC' refers to the data rate within the optical medium whilst 'STS' strictly refers to electrical transmission.

ATM and the ISO model for OSI

ATM operates at the Data Link and Physical Layers of the ISO model for OSI. Some higher level activities (such as routing) are also provided by ATM. Higher level protocols are added to improve reliability and integrity. ATM's Physical Layer provides the same service as the Physical Layer in the ISO model but transmits cells rather than individual bits. The ATM Layer sits above the Physical Layer and is equivalent to the ISO model's Data Link Layer. The ATM layer is responsible for reading and interpreting headers and routing information as well as for packaging the user's data into 48 byte payloads for transmission. The ATM Adaptation Layer also accepts the received 48 byte payloads and reassembles them into data.

10 Transmission protocols

In order to successfully transmit data from one location to another it is necessary to define a set of rules, known as protocols. Typically, these rules or protocols will say what should happens when a terminal is ready to transmit data, the format that the data should be in (e.g. packet definition), how the data is checked for errors and what happens if any are detected.

Flow control

Flow control is required in a data communications system in order to:
(a) ensure that transmission rates match the processing capabilities at each end of the link
(b) ensure that the capacity of buffer storage is not exceeded by the volume of incoming data.

Various protocols are in common use including XMODEM, YMODEM, and Kermit. Many communications software packages support several of these protocols and allow users to select that which is employed.

X-ON/X-OFF flow control

X-ON/X-OFF flow control is commonly used for serial data communications in conjunction with peripherals such as modems and serial printers. To stop a host from sending, the receiving device (peripheral) sends an X-OFF code. The host then waits until the receiving device generates an X-ON code before transmitting further serial data. X-OFF is equivalent to CTRL-S (ASCII 13 hexadecimal) whilst X-ON is equivalent to CTRL-Q (ASCII 11 hexadecimal).

XMODEM protocol

Definitions

	Meaning	Hexadecimal
SOH	start of heading	01
EOT	end of transmission	04
ACK	acknowledge	06
NAK	negative acknowledge	15
CAN	cancel	18

Transmission medium level protocol

Asynchronous data transmission with eight data bits, no parity bit, and one stop bit. The protocol imposes no restrictions on the contents of the data being transmitted. No control characters are looked for in the 128-byte data message blocks. Data may be transmitted in any form (binary, ASCII, etc.). The protocol may be used in a 7-bit environment for the transmission of ASCII-encoded data.

To maintain compatibility with CP/M file structure and hence allow ASCII files to be transferred to and from CP/M systems, the following recommendations have been made:

(a) ASCII tab characters (09 hexadecimal) should be set every eight character positions
(b) lines should be terminated by a CR-LF combination (0D hexadecimal followed by 0A hexadecimal)
(c) end-of-file should be indicated by CTRL-Z (1A hexadecimal)
(d) variable length data is divided into 128-byte blocks for transmission purposes
(e) if the data ends on a 128-byte boundary (ie, a CR-LF combination occurs in the 127th and 128th byte positions of a block) a subsequent block containing CTRL-Z should preferably be appended in order to indicate the end-of-file (EOF)
(f) the last block transmitted is not shortened (ie, all blocks have a length of 128 bytes – there is no *short block*).

Character

File level protocol

The XMODEM file level protocol involves the following considerations:

(a) all errors are retried ten times
(b) some versions of the protocol use CAN (CTRL-X) to abort transmission. Unfortunately, such a scheme can result in premature termination of a transmission due to corruption of data bytes (which may be falsely read as CAN)
(c) the receiver will *timeout* after ten seconds and then send a NAK character. The timeout sequence should be repeated every ten seconds until the transmitter is ready to resume the transmission of data. When receiving a block, the time out is reduced to one second for each character.

184

The following example shows how the XMODEM protocol provides for error recovery:

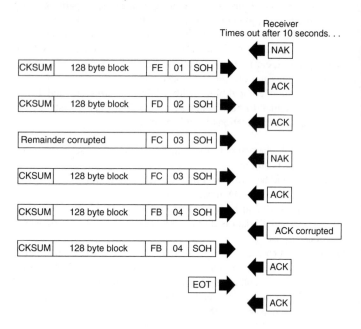

Message block level protocol

Each block transferred comprises:

SOH	blk#	255-blk#	128 data bytes	cksum

where:

SOH   is the start of heading character (01 hexadecimal)
blk#   is the block number (in pure binary) which starts at 1. The block number is incremented for each 128-byte block transmitted and wraps round from 255 to 0 (not from 255 to 1)
255-blk#   is the one's complement of the block number
cksum   is the sum of the data bytes within the 128-byte block (any carry is discarded).

Notes:

1. All single byte values are given in hexadecimal
2. SOH = 01, EOT = 04, ACK = 06, NAK = 15 (all hexadecimal)

XMODEM/CRC protocol

An improved version of XMODEM protocol exists in which the simple single byte checksum is replaced by a two-byte *cyclic redundancy check* (CRC) character. The CRC provides a more robust form of error detection

SOH	blk#	255-blk#	128 data bytes	CRC hi	CRC lo

where:

SOH	is the start of heading character (01 hexadecimal)
blk#	is the block number (in pure binary) which starts at 1. The block number is incremented for each 128-byte block transmitted and wraps round from 255 to 0 (not from 255 to 1)
255-blk#	is the one's complement of the block number
CRC hi	is the high byte of the CRC
CRC lo	is the low byte of the CRC.

The sixteen bits of the CRC are considered to be the coefficients of a polynomial. The 128-byte value of the data block is first multiplied by x^{16} and then divided by the generator polynomial $(x^{16} + x^{12} + x^5 + 1)$ using modulo-2 arithmetic $(x = 2)$. The remainder of the division is the desired CRC which is then appended to the block. The CRC calculation is repeated at the receiving end, dividing the 130-byte value formed from the 128-byte data block and two-byte CRC. If anything other than a zero results as the remainder generated by this calculation, an error must have occurred in which case a NAK will be generated in order to signal the need for retransmission of the block.

The file level protocol of XMODEM/CRC is similar to that used in the basic XMODEM specification with the exception that a C (43 hexadecimal) is initially transmitted by the receiver. This character is sent in place of the initial NAK. If the sender is set up to accept the modified protocol, it will respond by sending the first message block just as if a NAK had been received. If, however, it is not set up for the modified protocol, the sender will ignore the character. The receiver will then wait for 3 seconds and if no SOH character is detected, it will assume that XMODEM/CRC protocol is not available and will resume the data transfer by sending a NAK and adopting XMODEM protocol with a checksum for error detection.

The following examples show how the XMODEM/CRC protocol provides data transfer:

(a) With a sender set up for XMODEM/CRC

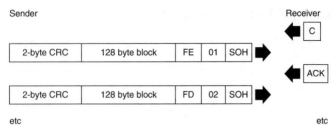

Notes:

1. All single byte values are given in hexadecimal
2. C = 43, SOH = 01 and ACK = 06 (all hexadecimal)

(b) with a sender not set up for XMODEM/CRC

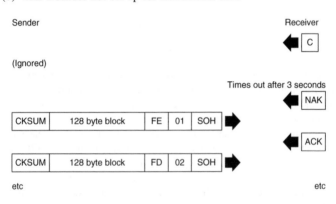

Notes:

1. All single byte values are given in hexadecimal
2. C = 43, SOH = 01, NAK = 15, ACK = 06 (all hexadecimal)

ITU X.25

ITU X.25 is a major protocol standard which has gained much support amongst computer and networks vendors alike. X.25 originated in 1976 (before the emergence of the ISO model for OSI) and thus it is perhaps not surprising that it does not conform exactly to this widely accepted model. In 1983, the US Government adopted a subset of X.25 for use by federal departments and agencies. This joint standard appears in Federal Information Processing Standard (FIPS) 100/Federal Standard (FED-STD) 1041.

X.25 corresponds to the lower layers of the ISO model (with some overlap) as shown below:

ISO layer	X.25
7 Application	
6 Presentation	
5 Session	
4 Transport	
3 Network	3 Packet level
2 Data link	2 Link level
1 Physical	1 Physical level

X.25 is best described as a *packet mode interface protocol* which also offers some end-to-end properties. X.25 has the following major characteristics:

Physical level

Transmission rates: 2.4 K, 4.8 K and 9.6 K bps
Interface requirements: RS-232, X21, RS-449

Link level

Procedure: linked access protocol (LAP) and link access protocol balanced (LAPB)

Maximum outstanding data frames: 7
Maximum number of bits per information frame: 164 octets

Packet level

Services: virtual call and permanent virtual circuit

Packet types: all basic types plus diagnostic packets

User data-field length: integral number of octets
Packet sequence numbering: modulo-8
Delivery confirmation: supported by all DCE (DTE need not employ the delivery confirmation bit when sending to the DCE)

Fast select: a DCE should implement fast select (a DTE need not employ fast select when sending to a DCE).

X.25 packet format

The general format of an X.25 packet is shown below:

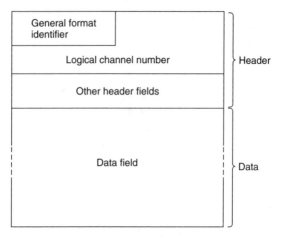

The general format identifier occupies the four most significant bits of the first octet and the logical channel number comprises group and channel numbers in the four least significant bits of the first octet and eight bits of the second octet respectively. The remainder of the header contains various items of information depending upon the type of packet.

X.25 control packet format

0 0 0 1	Group number	
Channel number		
Command type		1
Calling DTE address length	Called DTE address length	
DTE address		
		0 0 0 0
0 0	Facility length	
Facilities		
User data		

X.25 data packet format

Q	D	01	Group number	
Channel number				
Ack. number		M	Sequence number	O
User data				

Notes:

1. The Q bit (MSB of the first octet) provides a mechanism by which the user may operate two streams of data across a single virtual circuit. If this facility is not used, the Q bit defaults to 0. The Q bit is set to 1 when messages are sent to the packet assembly/disassembly (PAD) device. Packets destined for a terminal (or assembled by a PAD from a terminal) have the Q bit reset (ie, 0).

2. The D bit is set to logic 1 in order to indicate that immediate confirmation of packet receipt is required (rather than waiting until a window is full). In the call set-up routine, the D bit may be set by the sender to ascertain whether the receiver can support this feature. The response is indicated by the state of the D bit within the call accept packet.

Frame relay

Frame relay provides a packet switching network is a physical layer using a data link protocol. The physical layer is often narrow-band or broadband ISDN (ATM) using frame relay switching, although it is possible (but less efficient) to use time division multiplexers and packet switches.

The link access protocol, used by X.25 networks (LAPB) has been adapted for frame relay using ISDN networks as LAPD. This protocol provides congestion control and error detection, and is used to control closed user groups. The LAPD frame comprises: one flag byte, two address bytes, one control byte, payload data, followed by two separate cyclic redundancy check bytes (CRC1 and CRC2).

As a packet traverses a frame relay network, error checking is carried out at the frame relay switch, using the CRC bytes at the end of the LAPD packet. If an error is found, the packet is discarded. It is up to higher protocols to request retransmission of any packets that have not been received, and this is carried out at the end receiving station (not at the switch). Requests for retransmission are not carried out at the switch because this would reduce the switching speed −since ATM is very reliable the number of retransmission requests is low.

High-level data link control

High-level data link control (HDLC) is a synchronous communication protocol on which many of today's LAN protocols are based. HDLC is a *bit-oriented protocol*; the bit representation of the data in the form of characters, binary numbers, or decimal numbers, is contained wholly within the data field of a single frame.

HDLC provides three classes of procedure for network connection between adjacent nodes in a point-to-point communication system.

Asynchronous balanced mode

Asynchronous balanced mode (ABM) relates to full duplex communication between two nodes who are considered to be 'equal partners' in the data exchange. Both can initiate and terminate a connection and both can send data (without prior interrogation) on an established connection.

Normal response mode

Normal response mode (NRM) relates to communication between a control device (eg, a computer) and a number of secondary stations.

Asynchronous response mode

Asynchronous response mode (ARM) relates to communication in half-duplex mode between a primary station which sends out commands and data and a secondary station which returns responses.

HDLC frame structure

A number of fields are employed within an HDLC frame, including:

Flag

The flag is used for synchronisation and also indicates the start (*preamble*) and end (*postamble*) of the frame. The flag takes the form:

01111110, and this pattern is avoided in the data field by a technique known as *bit stuffing*.

Bit stuffing is the name given to a process in which the transmitter automatically inserts an extra 0 bit after each occurrence of five 1 bits in the data being transmitted. When the receiver detects a sequence of five 1 bits, it examines the next bit. If this bit is a 0, the receiver deletes it. However, if this bit is a 1, it indicates that the bit pattern must form part of preamble or postamble code.

Address

Identifies the sending or receiving station.

Control

The control frame is used to identify one of three different types of frame:

(a) information frame (the frame contains data)
(b) supervisory frame (the frame provides basic link control functions)
(c) un-numbered frame (the frame provides supplementary link control functions)

Data

The data frame contains the data to be transmitted.

Frame check sequence

The frame check sequence (FCS) comprises a 16-bit *cyclic redundancy check* (CRC) which is calculated from the contents of the address, control, and data fields.

HDLC frame structure

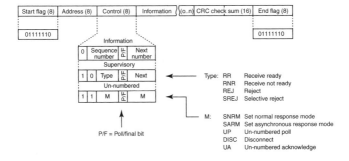

LAN software

The software required to operate a local area network successfully bridges the gaps between the session layer, presentation layer, and applications layer of the ISO model for OSI. In addition, software defined for use in conjunction with the lower layers of the model is concerned with the efficient transport of data within the LAN and for establishing the dialogue between users, servers, and other resources. Such software is typified by IBM's NETBIOS and the IBM PC LAN Support Program designed to support token ring networks and Xerox Network Services (XNS).

The relationship between the seven-layered ISO model and the categories of software present is shown below:

ISO layer	Software
7 Application	User application (eg, dBase IV, Word, WordPerfect, etc.)
6 Presentation	Network operating system and LAN utilities (eg, NetWare, etc.)
5 Session	Host operating system (eg, MS-DOS, OS/2, Unix, Xenix, etc.)
4 Transport	Network transport systems (eg, IBM NETBIOS, IBM PC LAN Support Program, Novell IPX, Xerox Network Service (XNS), Transport Control Protocol/Internet Protocol)
3 Network	(TCP/IP, etc.)
2 Data link	Link control/media access
1 Physical	LAN access/ signalling (eg, Token Ring, Ethernet, Arcnet, etc.)

LAN security and management

Intranets and firewalls

Internet technologies are also used for internal company networks, known as an intranet. An intranet is an internal company network based on the Internet protocol (IP) and makes use of WWW server and client technology, e-mail, etc. An intranet may also include links that cross the public Internet to connect together a company's different local area networks. The key benefits and opportunities of an intranet include better communications, internal e-mail, mailing lists, etc., and allows more effective publishing and distribution of information within an organisation (employee handbooks, quality management systems, etc.) which need version control.

Connection between the intranet and the Internet has security risks, including access by hackers or competitors to commercial information. Encryption can be used to make e-mail secure and restricting access to information can be achieved by installing a 'firewall' between the internal network and the Internet.

Firewalls are installed at the network interface (usually on a router) and are either packet screens or proxy servers. Packet screens examine each packet passing to and from the internal network. Packets are either allowed through or discarded depending on the rules applied to the firewall. Proxy servers control the type of services that may pass through the firewall, for example WWW or e-mail. Proxy servers are very generally secure.

Virtual private networks (VPNs)

Confidence in the security of IP networks is increasing and Intranets are now being provided on a shared IP infrastructure, on what is referred to as a virtual private network (VPN). Potentially, thousands of VPNs can be provided over a single shared high-capacity global IP network. VPNs can also support extranets, in which secure connectivity is extended to suppliers, customers or communities of interest over a common IP infrastructure.

Electronic commerce

Many companies now have on-line shops through which goods can be ordered. Travel, banking and insurance companies have been particularly active, but security is an issue. Cryptography provides the security necessary for privacy of communications and payments to be made over the Internet. Two forms of encryption are in common use: symmetric encryption and asymmetric encryption (also known as public-key cryptography). Both use cryptographic algorithms and keys to encode and decode data. The keys are parameters used in the mathematical encryption process.

In symmetric key systems, the sender and receiver must use identical keys. The security of symmetric key systems depends on keeping the key information private, therefore the key must be sent over a separate and secure path. This is a disadvantage for widespread use in electronic commerce. Symmetric encryption algorithms are much faster than public-key algorithms and the Digital Encryption Standard (DES) is widely used for this.

Public-key schemes use pairs of related keys. Each user has a private key, which is kept secret, and a public key, which is published and readily available to others. Encryption of a message with the recipient's public key provides confidentiality because the owner of the corresponding private key is the only person who can read it. This approach is often used as a secure way of exchanging symmetric keys. Encryption of a message with the sender's private key also provides a signature since the message can only be decrypted correctly using the sender's public key.

The overall security of key-based systems depends on the strength of the cryptographic algorithms and on the security of the keys. The strength of the encryption increases exponentially with key length.

Network computing

In network computing all software and data is stored remotely on servers in the network and downloaded to the user's PC as and when required. The main advantages of network computing are reduced version management, and rapid access to new applications and services.

Java

The basis of network computing is the Java programming language. The software required to execute Java is available within every WWW browser, which means that almost every networked computer is capable of running a Java program. Java offers animation and the full power of a computer programming language in the browser. Java applications can expect to be seen running independently of the browser.

Real-time services

Limits in processing speed and Internet bandwidth have restricted the development of real-time services. But recent developments in PC technology have enabled video compression within a realistic period. Competition in telecommunications has created considerable Internet bandwidth. The high speed processors and wide bandwidth have led to a rapid increase in the speed of real-time services development over the Internet, including Internet telephones, audio and video streaming applications.

The prospect of 'free' calls using Internet telephony has excited some enthusiasts, but the truth is the fundamental costs of Internet telephony and PSTN telephony are similar. However Internet telephony has yet to match the convenience, ease of use, call quality and customer service of the PSTN. However, Internet telephony is easily integrated with other computer applications and so is likely to continue to be used and to develop. A gateway is needed to provide PSTN interconnectivity for telephone calls between IP networks and the PSTN.

Voice over IP is offered on a 'best effort' basis and the quality of the call is dependent upon network congestion. Factors that affect call quality are packet loss and wide variations in network delay. The IETF is developing the integrated services architecture (ISA), which is a framework that aims to offer control over the quality of service provided by a network. ISA allows applications to prioritise their traffic and to request network resources to enable the priority system to work. Two major components of ISA are IP multicast and resource reservation protocol (RSVP).

The quality of service seen by the user depends on the quality of service provided by the network, the operating systems and the application protocols. The requirement for a reliable signal conflicts with the requirement for low delay. Retransmission of a packet found to contain errors will lead to increased end-to-end delay and increased jitter (variation in end-to-end delay).

The enhancement of the Internet and intranets to support real-time services is exciting, because it promises a future in which a single network can be used for voice, data and all other media. It also promises excellent computer/ telephony integration (CTI) and interactive entertainment.

Differentiated services

One approach to providing quality of service (QoS) in core IP networks is based on the IETF standard of differentiated services (Diffserv). Diffserv gives high priority to certain types of data; for example, gold service for delay-sensitive voice or video traffic, silver for medium-priority services such as e-mail, and bronze for low-priority data. Priorities can be identified in several ways, including the type of service bits in the IP packet header. One technique for giving different levels of priority is weighted random early discard (WRED), which selectively discards low priority packets at the edge of the network to protect the core from congestion. Another technique is class-based queuing (CBQ), in which bandwidth and delay limits are set. Packets are then processed through the router according to their class or priority.

Network operating systems

A network operating system (NOS) provides added value to an individual PC or work station by facilitating resource sharing and information transfer via the LAN. The NOS is thus crucial in determining the overall effectiveness of the system as well as the transparency of the network in terms of access to the communications, file, and print services offered by the network server.

Software architecture of a typical network work station (PC)

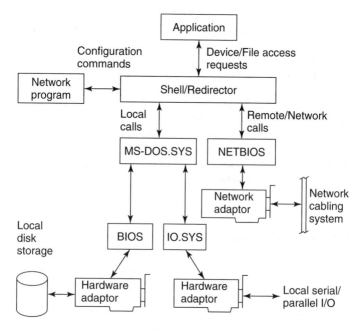

Software architecture of a typical file server

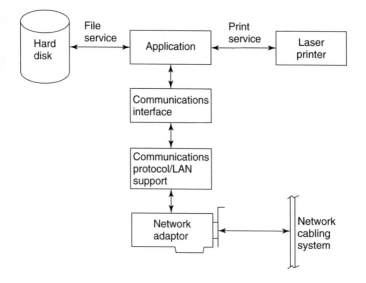

NOS facilities

In general, a network operating system should:

(a) provide access to files via the file server on a multitasking basis
(b) provide a user shell which, in conjunction with the host operating system (eg, MS-DOS), will redirect network file requests
(c) provide file and record locking
(d) include transaction support (read/modify/write)
(e) manage a print queue (normally at the file server)
(f) incorporate a significant element of fault tolerance (including redundant directory management, power supply monitoring, transaction tracking, etc.)
(g) Incorporate differing levels of security and/or access control
(h) provide network accounting facilities
(i) permit inter-networking via internal and/or external bridges (and asynchronous communications, where appropriate)
(j) incorporate message handling facilities for 'store and forward' communications.

Network protocols

SPX/IPX

Novell was the first vendor to introduce a network operating system (NOS) that would support true file-sharing. NetWare is the name of Novell's suite of network utilities that allows any PC platform to share files, CD-ROM drives, and printers across a LAN. The suite works with almost any popular LAN standard including Ethernet, Token Ring, and ARCnet.

Novell's SPX (Sequenced Packet Exchange) and IPX (Internet Packet Exchange), respectively, provide the Transport Layer and Network Layer of the ISO model for OSI. SPX/IPX is not a single protocol but a suite of standard procedures for connecting computers. IPX is responsible for addressing packets between NetWare nodes while SPX encapsulates IPX packets for reliable data transfer. Many applications, however, are able to monitor the reliability of data transfer themselves and they make use of IPX alone, thus reducing overhead and improving efficiency.

IPX is fast and efficient with the relatively small packets (eg, 512 bytes) found in a PC networking environment under DOS or Windows. In the context of wide area networks (WAN), small packets are inefficient due to the appreciable overhead that can be present when a large amount of data is to be transferred. Packet burst mode (first introduced in Net Ware 3.x) allows the 512 byte packet length to be exceeded. IPX packets of up to 64 kbyte can be supported in this mode. Burst mode can greatly improve the efficiency of an expensive long-distance low-speed circuit. The improvement becomes even more significant when large files are to be exchanged.

Internet protocol (IP)

The foundation of the Internet is the internet protocol (IP), which routes data packets across the network from one computer to another. IP is very simple because it focuses specifically on packet routing but does not guarantee packet delivery. Packets may be lost or corrupted, and a sequence of packets sent between two hosts may take different routes and may arrive in a different order. These issues are left for other protocols to handle.

IP packets are sent from one router to another towards their destination. Each router maintains routing tables, which determine the output interface on which a packet should be sent. When a packet arrives at a router, it is stored in a queue. When it reaches the front of the queue, the router reads the destination address in the header. This address is looked up in the routing table and then the router transmits the packet out of the appropriate interface port. Routing tables are regularly updated to take account of network failures or changes to the configuration of the network.

If the network is congested, the input queues at routers can become very long, which increases the packet transit delay. Under heavy congestion the queues may become full and packets are discarded. It is up to other protocols (such as TCP) to recognise a missing packet and to request retransmission.

The military origin of IP has helped it to become the *de facto* internetworking standard. The lack of delivery-guarantee makes it simple and efficient, because not all applications require these guarantees. Also, its ability to adapt to network failures and configuration changes has made IP particularly robust. Its ability to adapt has enabled it to cope with the exponential Internet growth and allows networks of many sizes and type to be easily connected.

IPv4

Currently, all IP networks use Version 4 of the Internet protocol. This was developed before it was realised how popular IP would become. The greatest limitation of IP version 4 (IPv4) is the size of the address space, which allows only a 32-bit address. This is not enough for everyone to be allocated a unique address. Hence IP addresses have become scarce and schemes have had to be developed to enable sufficient addresses. One technique is network address translation (NAT), which translates and isolates the Internet address so that previously used addresses can be re-used on the internal network side of the router.

The Internet Engineering Task Force (IETF) has been working to design the next generation of IP, IPv6 (the proposed IPv5 was rejected). A global experimental network has been built, called the 6bone, with the aim of evaluating early IPv6 implementations and identifying any problems. The IPv4 header contains 10 fields, a checksum, two addresses and some options, see diagram below. IPv4 addresses are based on a 32-bit word that, in theory, gives IPv4 over 4 billion addresses. The options field in the IP header is a disadvantage because it requires recalculation of a checksum for every possible IPv4 header type.

Ver	HL	TOS	Payload length	
Fragment identification			Flags	Fragment offset
Time-to-live		Protocol	Header checksum	
Source address				
Destination address				
Options (if any)				Padding
Payload (packet contents)				

IP header fields

VER (version) = 4 bits

The version field indicates the format of the header. In this case it is version 4.

HL (header length) = 4 bits

The header length is the length of the IP header in 32 bit words and points to the beginning of the payload. Note that the minimum value for a correct IPv4 header is 5, but this value is increased if the options field is used. The maximum header length is 60 octets (including options).

TOS (type of service) = 8 bits

The type of service provides an indication of the quality of service desired. These parameters are used when transmitting a packet through a particular network and are a guide to the selection of the actual service. Some networks treat high precedence traffic as more important than other traffic. During periods of high load, only traffic above a certain level of precedence is accepted. The choice is a trade-off between low-delay, high-reliability and high-throughput.

Bits 0–2 determine the precedence, given by the following values:

111 – Network Control
110 – Inter-network Control
101 – CRITIC/ECP
100 – Flash Override
011 – Flash
010 – Immediate
001 – Priority
000 – Routine

Bit 3 determines the delay (0 = normal delay, 1 = low delay). Bit 4 determines the throughput (0 = normal, 1 = high) Bit 5 determines the reliability (0 = normal, 1 = high) and Bits 6–7 are reserved for future use.

Total length = 16 bits

Total length is the length of the packet, measured in octets, including the header and payload. This field allows the length of a packet to be up to 65,535 octets. Such long packets are impractical for most hosts and networks; Ethernet based networks limit packets to 1500 octets.

All hosts must be prepared to accept packets of up to 576 octets (whether they arrive whole or in fragments). It is recommended that hosts only send packets larger than 576 octets if they have checked that the destination is prepared to accept them. The number 576 is selected to allow a reasonable sized data block to be transmitted in addition to the required header information. For example, this size allows a data block of 512 octets plus 64 header octets to fit in a packet. The maximal IP header is 60 octets, but is typically 20 octets, thus allowing a margin for headers of higher level protocols.

Fragment identification = 16 bits

An identifying value assigned by the sender to aid in assembling the fragments of a packet.

Flags = 3 bits

Various control flags. Bit 0 is reserved and must have a value of zero. Bit 1 (DF) (0 = Fragmentation allowed, 1 = do not fragment). Bit 2 (MF) (0 = last fragment, 1 = more fragments).

Fragment offset = 13 bits

This field indicates where in the packet this fragment belongs. The fragment offset is measured in units of 8 octets (64 bits). The first fragment has offset zero.

Time-to-live = 8 bits

This field indicates the maximum time the packet is allowed to remain within the Internet. If this field contains the value zero, then the packet must be destroyed. The value in this field is reduced by at least one during IP header processing by a router. The 'time-to-live' is measured in units of seconds, but if the router process the packet in less than one second the T-T-L will be shorter than expected. In fact, the T-T-L must be thought of as an upper limit on the time a packet may exist. The aim is to discard undeliverable packets.

Protocol = 8 bits

This field indicates the next higher level protocol used in the payload portion of the IP packet. The values for various protocols are specified in 'Assigned Numbers', and examples are: 06 (hex) = TCP, 11 (hex) = UDP and 58 (hex) = IGRP (Cisco proprietary).

Header checksum = 16 bits

This is a checksum on the header only. Since some header fields change (e.g., time-to-live), this is computed and verified at each point that the IP header is processed.

The header is broken down into 16-bit words, which are then summed using one's complement arithmetic with any carry added in. A one's complement of the result gives the checksum. For the purpose of this computation, zero value is used for the checksum field. This checksum is simple to compute and has been found to be adequate.

Source address = 32 bits

The source address.

Destination address = 32 bits

The destination address.

Options: variable length

The options field is used mainly for IP packet tracing, time-stamping and security. In some cases the security option may be required in all packets. The option field is variable in length; there may be zero or more options.

There are two formats for options. The first is a single octet of option-type. The second format is a sub-packet containing an option-type octet, an option-length octet and option data of variable length. The option-length given in the second octet is the length of the whole sub-packet.

The option-type octet is viewed as having 3 fields: 1 bit is the copied flag, 2 bits are the option class and 5 bits are the option number. The copied flag indicates that this option is copied into all fragments on fragmentation (0 = not copied, 1 = copied).

The option classes that are used are 0 for control and 2 for debugging and measurement. Option classes 1 and 3 are reserved for future use.

The following IP options are defined:

Class	Number	Length	Description
0	0	–	End of Option list. This option occupies only one octet; it has no length octet.
0	1	–	No Operation. This option occupies only one octet; it has no length octet.
0	2	11	Security. Used to carry security, restrictions compatible with DOD requirements.
0	3	variable	Loose Source Routing. Used to route the IP packet based on host supplied information.
0	9	variable	Strict Source Routing. Used to route the IP packet based on host supplied information.
0	7	variable	Record Route. Used to trace the route an IP packet takes.
0	8	4	Stream ID. Used to carry the stream identifier.
2	4	variable	Internet Timestamp.

Specific option definitions

Padding: between 1 and 3 octets long

The IP header padding is used to ensure that the IP header ends on a 32-bit boundary. The value of the padding is zero.

IPv6

The IPv6 header has six fields and two addresses, see diagram below. The address space in the IP header is 128 bits and gives the possibility of 3.4×10^{38} addresses. In IPv6 the options field is placed after the IPv6 header and before the data. Thus the IPv6 header has a fixed size and hence a checksum field is not needed. This also makes IPv6 headers suitable for processing by hardware, owing to their fixed size.

Ver	Traffic class	Flow label	
Payload length		Next header	Hop limit
Source address			
Destination address			
Payload (packet contents)			

The IPv6 header has the following fields:

1. Version, a 4-bit field that identifies the IP version used.
2. Traffic class, an 8-bit field that is used for distinguishing between different classes or priorities of IPv6 packets. These bits may be used to offer various forms of differentiated service (i.e. priority) for IP packets.
3. Flow label, a 20-bit field that is used to identify a flow of IP packets between a particular source and destination.
4. Payload length, a 16-bit field that gives the length of the IPv6 payload.

5. Next header, an 8-bit field that identifies the type of header immediately following the IPv6 header, for example, UDP or TCP. This
is the same as used in IPv4.

6. Hop limit, an 8-bit field whose decimal value is decreased by one
each time the packet passes through a router. The IP packet is
discarded if this value reaches zero.

7. Source address, a 128-bit field that represents the source address
of the IPv6 packet.

8. Destination address, a 128-bit field that represents the destination
address of the IPv6 packet.

Each IPv6 address is represented by eight groups of 16-bit numbers,
separated by colons. Each number is displayed as four hexadecimal
figures. An example of an IPv6 address is 16EE:3B40:0000:0000:
000C:03E9:7ADE:9BB7. Where there are zeroes in a hexadecimal
block, a double-colon can replace them. Thus the same IPv6 address
can be represented as: 16EE:3B40::C:3E9:7ADE:9BB7.

IPv6 has been designed to support 'real-time' traffic, such as voice
and video. It does this using the traffic class field and the flow label
field, and these are used to minimise the delay and delay variation of
the IP packets. The traffic class field is assigned different values (1
to 8) dependent upon the type of IP data, typically video could be
assigned a low value and e-mail a high value, the lower values giving
greater priority. The flow label field is used to identify a stream of
IPv6 packets that each have a particular destination address and a
particular source address. A router will be able to identify packets
from their flow label and simply forward them without having to read
the address fields.

It is probable that an IPv6 Internet will be operated in parallel with
the existing IPv4 Internet. Host computers will contain both the IPv4
and IPv6 stack, and applications using the Internet will determine the
IP protocol version from information provided by the domain name
service (DNS).

Transport protocols

Transport protocols are used together with IP in order to provide error
checking, error correction or recovery. The transport frame (including
data and transport header) is placed inside the Internet packet and
forms the data part of an IP packet.

There are two main transport protocols in use on the Internet – UDP and TCP. User packet protocol (UDP) is intended for sending messages without guarantee of arrival and without notifying the sender of successful or failed delivery. It is very simple. In addition to the data carried within the packet, information is supplied about the application being used and a checksum is transmitted to indicate whether any data has been corrupted in transit. In contrast to UDP, the TCP protocol is designed to handle all types of network failure. If packets are lost or corrupted then TCP arranges for them to be re-transmitted. If packets arrive out of order then TCP will re-order them. If packets are repeatedly lost, the TCP source will assume that the network is congested and reduce its transmission rate.

Ports

At any one time, a terminal connected to an IP network may be interacting with other hosts and a number of different applications may be involved. Port numbers are used to identify for which application and for which activity a packet is intended. Each UDP or TCP packet carries two port numbers; one port number identifies the server application and the other port number is selected by the client to distinguish the particular activity. Commonly used ports, which are used with applications like Telnet and FTP, are defined in the assigned numbers RFC. An RFC is a request for comments and is part of the standardisation process used by the IETF.

Telnet

Telnet is a very basic Internet function that allows a user to access a server remotely. This is known as remote terminal access and works by carrying ASCII text (typed by the user) to the remote server and then returning the output from the application on the remote server to the user.

File transfer protocol (FTP)

FTP provides a basic service for the reliable transfer of files from one machine to another and allows the user to establish a control connection between their client and the server. This connection can be used to navigate through the server's directory structure and request the transmission of files. A separate data connection is set up to transmit the files.

Domain name service (DNS)

The domain name service provides name/address translation for all objects on the Internet. Every computer (and every router) on the Internet has an individual name written by concatenating two or more names using '.' as a separator, e.g. ieee.org. In this name the top level domain is 'org' and the IEEE owns the sub-domain (or name-space) 'ieee'.

The owner of a 'name space' must run a DNS server, which presents its address to the DNS server at the next level up. Applications such as FTP, SMTP, Telnet and WWW send a request to the local DNS server, which responds by producing the answer itself or with the address of a DNS server that can provide the answer. If the local DNS server responds with the answer, then this may have come either from its own look-up tables or from another DNS server (that it approached on the application's behalf).

E-mail

Electronic mail or e-mail is the electronic equivalent of the traditional postal service. E-mail is sent from a mail 'client' (a program which runs on the user's machine) to the destination mailbox using the simple mail transfer protocol (SMTP). Other protocols such as POP3 or IMAP4 are used for checking and retrieving mail from a mailbox. SMTP routes emails from the mail agent to the destination mailbox via a number of mail handlers. Mail handlers behave like conventional sorting offices; that is, they sort the mail and pass it on to the next mail handler.

An SMTP mail address is based on an Internet domain name and takes the form user@domain, e.g. steve.winder@ieee.org. In order to route the message to an e-mail address of this form, the mail agent depends on DNS to translate the domain part of the e-mail address (the part after '@') into an IP address.

An email can be sent to a single recipient or a list of people with equal ease. E-mail lists are a simple way of informing a group of people. Unfortunately, e-mail lists can distribute junk mail with equal ease and can be used to spread computer viruses (as attachments).

World wide web

The application that has caused the growth in Internet use is the World Wide Web (WWW). The WWW makes publishing, display and retrieval of all kinds of data very easy and now accounts for most of the traffic on the Internet. Documents on the WWW are usually written in hypertext mark-up language (HTML). HTML is a program listing that describes the layout of the page and the components on it, such as the text and images. HTML pages are stored at servers and the information is accessed using a client application known as a browser; the most popular browsers are Netscape Navigator and Microsoft's Internet Explorer.

A Web page is identified by a uniform resource locator (URL). The browser and web-server usually communicate using hyper-text transfer protocol (HTTP), but HTTP is not the only protocol supported by browsers (for example, FTP and Telnet are also supported). Therefore, the URL must specify the file to be retrieved and the protocol to be used. In general a URL takes the form:

<protocol>: // <hostname>: <port> <directory> <filename>

WWW pages can use a wide variety of data types including text, image, animation, audio or video. The web-server specifies the type of data using a multipurpose Internet mail extension (MIME), which was originally designed for sending multimedia e-mail. The browser can handle some MIME content types, for example text/plain, text/HTML or image/GIF. Other formats such as application/postscript, audio/MPEG or video/MPEG are normally handled by a separate application, known as a 'helper' or a 'plug-in'.

The components (such as text, pictures, sound, etc.) of an HTML-based document may be distributed across a number of servers. Only the HTML program/framework has to exist on the web-server. The program includes both a URL and fetch instruction for each component required on the page, for processing by the WWW client. The content presented at a WWW client may be from stored files or from dynamic generation when the program is run on the server. The common gateway interface (CGI) is used to communicate to the web-server and is commonly used to perform searches.

Wireless application protocol (WAP)

Wireless application protocol (WAP) provides the equivalent of HTML on mobile terminals. Originally it was intended to provide a translation of wireless mark-up language (WML), but this failed and hence

WAP-enabled pages have to be published on the Internet to enable mobile terminals to view them.

TCP/IP and the ISO model for OSI

TCP operates at the level of the Transport Layer while IP provides the Network Layer of the OSI seven-layer model. Higher layers of the ISO model contain the utilities that constitute the full TCP/IP suite.

Like TCP, the User Datagram Protocol (UDP) also appears in the Transport Layer. UDP performs a similar function to TCP but does not retransmit data when errors occur. UDP can thus be considered less reliable but this problem is less significant when individual applications incorporate their own error checking routines.

FTP (File Transfer Protocol) occupies the Session Layer and part of the Presentation Layer while the Data Link and Physical Layers are typically provided by an Ethernet LAN. This scheme has the advantage that Ethernet adaptors are available for almost every type of machine. However, if national or international network access is required, X.25 packet switching can be used instead of a LAN standard.

TCP/IP thus provides both a wide area network (WAN) protocol and facilities for network resource management. TCP/IP is now widely available (often with a Microsoft Windows front-end). TCP/IP is also currently the most popular protocol for PC-to-Unix connectivity.

Routing protocols

Multi-protocol label switching (MPLS) is being developed by the Internet Engineering Task Force (IETF) and is based on Cisco protocols. It is used to efficiently route a packet through an MPLS enabled network. When a packet enter the edge of this network, a router reads the destination of the packet and request a route through the network to the point where the packet leaves the network.

The first router (edge label switching router) adds a short MPLS address label for use across one link within the network. The receiving router strips this address off and applies a new address for the next leg of the journey. At the final router, the MPLS address label is removed and the packet forwarded.

Although it appears to be adding complexity, this is not so. The first router must identify and read the address label of the incoming packet, which could be based on one of a number of protocols. The MPLS label is short and common to all types of packet, so reading the address is simple and quick. The main delay is in reading the address as the packet enters the network, once this is done the packet can be routed quickly. Without MPLS, each router would have to individually read the address for each type of packet, which would be slow.

Similar protocols are border gateway protocol (BGP), which is for inter-domain routing, and interior gateway protocol (IGP) that is used for routing within a single domain. Resource reservation protocol (RSVP) is used with label switching routers to reserve resources in order to guarantee a certain quality of service.

SNMP

Simple Network Management Protocol (SNMP) uses UDP as its transport mechanism. SNMP uses 'managers' and 'agents' rather than using a client and server as in overall TCP/IP. A manager communicates across the network while an agent provides information about a specific device. As well as being used extensively on the Internet, SNMP is also widely used in many commercial products.

ICMP

Internet Control Message Protocol (ICMP) checks on the status of devices on the network. Messages are generated when problems are found. ICMP usually operates in conjunction with IP.

NFS

Network File Server (NFS) is a set of protocols which allows multiple hosts to access files transparently from each other. This is achieved by using a distributed file system scheme.

RPC

Remote Procedure Calls (RPC) are functions that allow applications to communicate with a server. RPCs supply functions, variables and return values to support a client/server architecture.

Comparison of network protocols

	Novell NetWare	Microsoft Windows NT	Banyan VINES	TCP/IP	
Application layer	NetWare shell	Redirector	Redirector	NFS · NFS · TELNET · SNMP	Layer 7
Presentation layer	NCP	SMB	RPC	NFS · SMB	Layer 6
Session layer	NetBIOS	IPX or TCP	NetBIOS	NFS · RFS · SMTP · FTP · NFS	Layer 5
Transport layer	SPX	IPX or TCP	VIPC	TCP · UDP	Layer 4
Network layer	IPX		VIP	IP · X.25	Layer 3
Data link layer	ODI	NDIS	NDIS	MAC IEEE 802.3 · MAC IEEE 802.5 · LAPB	Layer 2
Physical layer	RS-232 or RS-449	RS-449 or 10BaseT	10BaseT	IEEE 802.3 or Ethernet · IEEE 802.5	Layer 1

NB: Some protocols do not fit neatly into the ISO model and thus the relationship between protocols should only be taken as approximate.

11 Reference information

ITU-T recommendations

The International Telecommunications Union – Telecommunications (ITU-T) produces internationally agreed standards for telecommunications. These standards appear as a number of recommendations which cover telecommunications apparatus and the transmission of both analog and digital signals.

The International Telecommunications Union (ITU) is the parent body for the ITU-T and is itself organised by and responsible to the United Nations.

The major ITU-T recommendations are organised into series which deal with data transmission over telephone circuits (V-series recommendations), data networks (X-series recommendations), digital networks (G-series recommendations), and integrated services digital networks (I-series recommendations).

ITU-T G-series recommendations

The following ITU-T G-series recommendations relate to transmission systems and multiplexing equipment characteristics of digital networks:

G.701	General structure of the G.700, G.800 recommendations
G.702	Terminology used for pulse code modulation (PCM) and digital transmission
G.703	General aspects of interfaces
G.704	Maintenance of digital networks
G.705	Integrated services digital networks (ISDN)
G.706	Frame alignment and cyclic redundancy check (CRC) procedures relating to G.704
G.707	Network node interface for the synchronous digital hierarchy (SDH)
G.711	Pulse code modulation (PCM) of voice frequencies
G.712	Performance characteristics of PCM channels at audio frequencies
G.720	Characterization of low-rate digital voice coder performance with non-voice signals
G.721	Hypothetical reference digital paths
G.722	7 kHz audio-coding within 64 kbit/s

G.743	Second order digital multiplex equipment operating at 6312 kbps and using positive justification
G.744	Second order PCM multiplex equipment operating at 8448 kbps
G.745	Second order digital multiplex equipment operating at 8448 kbps and using positive/zero/negative justification
G.746	Frame structure for use with digital exchanges at 8448 kbps
G.751	Digital multiplex equipment operating at third order bit rates of 34 368 kbps and fourth order bit rates of 139 264 kbps and using positive justification
G.752	Characteristics of digital multiplex equipment based on second order bit rates of 6312 kbps and using positive justification
G.753	Third order digital multiplex equipment operating at 34 368 kbps and using positive/zero/negative justification
G.754	Fourth order digital multiplex equipment operating at 139 264 kbps and using positive/zero.negative justification
G.755	Digital multiplex equipment operating at 139 264 kbit/s and multiplexing three tributaries at 44 736 kbit/s
G.763	Digital circuit multiplication equipment using ADPCM (G.726) and digital speech interpolation
G.764	Voice packetisation protocols
G.765	Packet circuit multiplication equipment
G.766	Facsimile demodulation/remodulation for digital circuit multiplication equipment
G.767	Digital circuit multiplication equipment using 16 kbit/s LD-CELP
G.780	Vocabulary of terms for synchronous digital hierarchy (SDH)
G.783	Characteristics of synchronous digital hierarchy (SDH) equipment
G.784	Synchronous digital hierarchy (SDH) management

ITU-T I-series recommendations

The following ITU-T I-series recommendations relate to integrated services digital networks (ISDNs):

I.110	General structure of the I-series recommendations
I.111	Relationship with other recommendations relevant to ISDN
I.112	Vocabulary of terms for ISDN
I.113	Vocabulary of terms for broadband aspects of ISDN
I.114	Vocabulary of terms for universal personal telecommunication

I.420	Basic access user-network interface
I.421	Primary rate user-network interface
I.430	Basic user-network interface – layer 1 specification
I.431	Primary rate user-network interface – layer 1 specification
I.432.x	Higher rate user-network interface
I.440	ISDN user-network interface: layer 2 – general aspects (Q.920)
I.441	ISDN user-network interface data link specification (Q.921)
I.450	ISDN layer 3 specification (Q.931)
I.460	Multiplexing, rate adaption and support of existing interfaces
I.461	Support of X.21 and X.21 bis based DTEs by an ISDN (X.30)
I.462	Support of packet mode terminal equipment by an ISDN (X.31)
I.463	Support of DTEs with V-series type interfaces by an ISDN (V.110)
I.464	Rate adaption multiplexing and support of existing interfaces for restricted 64 kbps transfer capability
I.465	Support by an ISDN of data terminal equipment with V-series interfaces (V.120)
I.731	Types and general characteristics of ATM equipment
I.732	Functional characteristics of ATM equipment

ITU-T V-series recommendations

The following ITU-T V-series recommendations cover data transmission over the telephone network:

V.1	Equivalence between binary notation symbols and the significant conditions of a two condition code
V.2	Power levels for data transmission over telephone lines
V.3	International alphabet no. 5
V.4	General structure of signals of international alphabet no. 5 code for data transmission over public telephone networks
V.5	Standards of modulation rates and data signalling rates for synchronous data transmission in the general switched network
V.6	Standards of modulation rates and data signalling rates for synchronous data transmission on leased telephone-type circuits
V.7	Definitions of terms concerning data transmission over the telephone network

V.8	Procedures for starting sessions of data transmission over the public switched telephone network
V.10	Electrical characteristics for unbalanced double-current interchange circuits for general use with integrated circuit equipment in the field of data communications (RS-423)
V.11	Electrical characteristics for balanced double-current interchange circuits for general use with integrated circuit equipment in the field of data communications (RS-422)
V.12	Electrical characteristics for balanced double-current interchange circuits for interfaces with data signalling rates up to 52 Mbit/s
V.13	Answerback unit simulator
V.14	Transmission of start-stop characters over synchronous bearer channels
V.15	Use of acoustic coupling for data transmission
V.16	Recommendations for modems for the transmission of medical dialog data
V.17	A 2-wire modem for facsimile applications with rates up to 14 400 bit/s
V.18	Requirements for DCEs operating in the text telephone mode
V.19	Modems for parallel data transmission using signalling frequencies
V.20	Parallel data transmission modems standardised for universal use in the general switched network
V.21	200 bps modem standardised for use in the switched telephone network
V.22	1200 bps full-duplex 2-wire modem standardised for use in the general switched telephone network and on leased lines
V.22 bis	2400 bps full-duplex 2-wire modem using frequency division techniques standardised for use in the general switched telephone network
V.23	600/1200 bps modem standardised for use in the general switched telephone network
V.24	List of definitions for interchange circuits between data terminal equipment and data circuit-terminating equipment (RS-232C)
V.25	Automatic calling and/or answering equipment on the general switched telephone network including disabling echo-suppressors on manually established calls

V.25 bis	Automatic calling and/or answering equipment on the general switched telephone network using the 100 series interchange circuits
V.26	2400 bps modem for use on 4-wire leased point-to-point leased telephone circuits
V.26 bis	2400/1200 bps modem standardised for use in the general switched telephone network
V.26 ter	2400 bps duplex modem using echo cancellation standardised for use in the general switched telephone network and on point-to-point 2-wire leased telephone circuits
V.27	4800 bps modem with manual equaliser standardised for use on leased telephone circuits
V.27 bis	4800/2400 bps modem with automatic equaliser standardised for use on leased circuits
V.27 ter	4800/2400 bps modem with automatic equaliser standardised for use in the general switched telephone network
V.28	Electrical characteristics for unbalanced double-current interchange circuits
V.29	9600 bps modem standardised for use on leased circuits
V.31	Electrical characteristics for single-current interchange circuits controlled by contact closure
V.31 bis	Electrical characteristics for single-current interchange circuits using optocouplers
V.32	Duplex modems operating at data rates of up to 9600 bps standardised for use in the general switched telephone network and in 2-wire leased telephone circuits
V.33	Full duplex synchronous or asynchronous transmission at 14.4 kbps for use in the public telephone network
V.34	33 600 bit/s modem for use on the PSTN and on leased lines
V.35	Interface between DTE and DCE using electrical signals defined in V.11 (RS-449)
V.36	Modems for synchronous data transmission using 60–108 kHz group band circuits
V.37	Synchronous data transmission at data rates in excess of 72 kbps using 60–108 kHz group band circuits
V.38	A 48/56/64 kbit/s data circuit-terminating equipment standardized for use on digital point-to-point leased circuits
V.41	Code-independent error control system
V.42	Error-correcting procedures for DCEs using asynchronous-to-synchronous conversion

V.42 bis	Data compression procedures for data circuit-terminating equipment (DCE) using error correction procedures
V.43	Data flow control
V.44	Error correcting protocol used with V.92 modems
V.50	Standard limits for transmission quality of data transmission
V.51	Organisation of the maintenance of international telephone-type circuits used for data transmission
V.52	Characteristics of distortion and error rate measuring apparatus for data transmission
V.53	Limits for the maintenance of telephone-type circuits used for data transmission
V.54	Loop test devices for modems
V.55	Specification for an impulsive noise measuring instrument for telephone-type circuits
V.56	Comparative tests for modems for use over telephone-type circuits
V.57	Comprehensive data test set for high signalling rates
V.90	A digital modem and analogue modem pair for use on the PSTN at data signalling rates of up to 56 000 bit/s downstream and up to 33 600 bit/s upstream
V.92	A digital modem and analogue modem pair for use on the PSTN at data signalling rates of up to 56 000 bit/s downstream and up to 44 000 bit/s upstream, using V.44 error correction
V.110	Support of DTEs with V-series type interfaces by an ISDN (I.463)

ITU-T X-series recommendations

The following ITU-T X-series recommendations cover public data networks:

X.1	International user classes of service in public data networks and ISDN
X.2	International user facilities in public data networks
X.3	Packet assembly/disassembly facility in a public data network
X.4	General structure of signals of international alphabet no. 5 code for data transmission over public data networks
X.5	Facsimile packet assembly/disassembly facility (FPAD) in a public data network
X.6	Multicast service definition

X.7	Technical characteristics of data transmission services
X.8	Multi-aspect PAD (MAP) framework and service definition
X.15	Definitions of terms concerning public data networks
X.20	Interface between data terminal equipment and data circuit-terminating equipment for start-stop transmission services on public data networks
X.20 bis	V21-compatible interface between data terminal equipment and data circuit-terminating equipment for start-stop transmission services on public data networks
X.21	General-purpose interface between data terminal equipment and data circuit-terminating equipment for synchronous operation on public data networks
X.21 bis	Use on public data networks of data terminal equipments which are designed for interfacing to synchronous V-series modems
X.22	Multiplex data terminal equipment/data circuit-terminating equipment for user classes 3–6
X.24	List of definitions of interchange circuits between data terminal equipment and data circuit-terminating equipment on public data networks
X.25	Interface between data terminal equipment and data circuit-terminating equipment for terminals operating in the packet mode on public data networks
X.26	Electrical characteristics for unbalanced double-current interchange circuits for general use in the field of data communications (identical to V.10)
X.27	Electrical characteristics for balanced double-current interchange circuits operating at data signalling rates up to 10 Mbit/s (identical to V.11)
X.28	Data terminal equipment/data circuit-terminating equipment interface for a start/stop mode data terminal equipment accessing the packet assembly/disassembly facility on a public data network situated in the same country
X.29	Procedures for exchange of control information and user data between a packet mode data circuit-terminating equipment and a packet assembly/disassembly facility
X.30	Support of X.21 and X.21 bis based data terminal equipment by an ISDN (I.461)
X.31	Support of packet mode terminal equipment by an ISDN (I.462)
X.32	Interface between data terminal equipment and data circuit-terminating equipment for terminals operating in packet

mode and accessing a packet switched public data network through a public switched network

X.33	Access to packet-switched data transmission services via frame relaying data transmission services
X.34	Access to packet-switched data transmission services via B-ISDN
X.35	Interface between a PSPDN and a private PSDN based on X.25.
X.36	DTE-DCE interface for public data networks providing frame relay
X.37	Encapsulation in X.25 packets of various protocols including frame relay
X.42	Procedures and methods for accessing a public data network from a DTE operating under control of a generalized polling protocol
X.45	DTE-DCE interface for packet mode terminals connected to public data networks
X.46	Access to FRDTS via B-ISDN
X.48	Procedures for basic multicast service using X.25
X.49	Procedures for extended multicast service using X.25
X.50	Fundamental parameters of a multiplexing scheme for the international interface between synchronous data networks
X.50 bis	Fundamental parameters of a 48 kbps user data signalling rate transmission scheme for the international interface between synchronous data networks
X.51	Fundamental parameters of a multiplexing scheme for the international interface between synchronous data networks using 10-bit envelope structure
X.51 bis	Fundamental parameters of a 48 kbps user data signalling rate transmission scheme for the international interface between synchronous data networks using 10-bit envelope structure
X.52	Method of encoding asynchronous signals into a synchronous user bearer
X.53	Number of channels on international multiplex links at 64 kbps
X.54	Allocations of channels on international multiplex links at 64 kbps
X.57	Method of transmitting a single lower speed data channel on a 64 kbit/s data stream
X.60	Common channel signalling for circuit-switched data applications

X.61	Signalling system no. 7 (data user part)
X.70	Terminal and transit control signalling system on international circuits between asynchronous data networks
X.71	Decentralised terminal and transit control signalling system on international circuits between synchronous data networks
X.75	Terminal and transit call control procedures and data transfer systems on international circuits between packet-switched data networks
X.76	Interface between public data networks providing the frame relay service
X.77	Interworking between PSPDNs via B-ISDN
X.80	Interworking of inter-exchange signalling systems for circuit switched data services
X.81	Interworking between an ISDN circuit-switched and a circuit-switched public data network (CSPDN)
X.82	Interworking between CSPDNs and PSPDNs based on Recommendation T.70
X.87	Principles and procedures for realisation of international test facilities and network utilities in public data networks
X.92	Hypothetical reference connections for public synchronous data networks
X.96	Call progress signals in public data networks
X.110	Routing principles for international public data services through switched public data networks of the same type
X.121	International numbering plan for public data networks
X.130	Provisional objectives for call set-up and clear-down times in public synchronous data networks (circuit-switching)
X.132	Provisional objectives for grade of service in international data communications over circuit-switched public data networks
X.150	Data terminal equipment and data circuit-terminating equipment test loops for public data networks
X.180	Administration arrangements for international closed user groups

The following ITU-T X-series recommendations relate to data communications networks for open system interconnection (OSI):

X.200	Reference model of OSI for ITU-T applications
X.210	OSI layer service definition conventions
X.213	Network service definition for OSI for ITU-T applications
X.214	Transport service definition for OSI for ITU-T applications

X.215	Session service definition for OSI for ITU-T applications
X.224	Transport protocol specification for OSI for ITU-T applications
X.225	Session protocol specification for OSI for ITU-T applications
X.244	Procedure for the exchange of protocol identical during virtual call establishment on packet-switched public data networks
X.250	Formal description techniques for data communications protocols and services
X.400	Message handling service for all test communications and electronic mail

Note: The words *bis* and *ter* refer to the second and third parts of the relevant ITU-T recommendation and these are usually concerned with enhancements to the original specification.

Powers of 2

n	2^n
0	1
1	2
2	4
3	8
4	16
5	32
6	64
7	128
8	256
9	512
10	1024
11	2048
12	4096
13	8192
14	16384
15	32768
16	65536
17	131072
18	262144
19	524288
20	1048576
21	2097152
22	4194304
23	8388608
24	16777216
25	33554432

n	2^n
26	67108864
27	134217728
28	268435456
29	536870912
30	1073741824
31	2147483648
32	4294967296

Power of 16

n	16^n
0	1
1	16
2	256
3	4096
4	65536
5	1048576
6	16777216
7	268435456
8	4294967296

Decimal, binary, hexadecimal and ASCII conversion table

Decimal	Binary	Hexadecimal	ASCII
0	00000000	00	NUL
1	00000001	01	SOH
2	00000010	02	STX
3	00000011	03	ETX
4	00000100	04	EOT
5	00000101	05	ENQ
6	00000110	06	ACK
7	00000111	07	BEL
8	00001000	08	BS
9	00001001	09	HT
10	00001010	0A	LF
11	00001011	0B	VT
12	00001100	0C	FF
13	00001101	0D	CR
14	00001110	0E	SO
15	00001111	0F	SI

Decimal	Binary	Hexadecimal	ASCII
16	00010000	10	DLE
17	00010001	11	DC1
18	00010010	12	DC2
19	00010011	13	DC3
20	00010100	14	DC4
21	00010101	15	NAK
22	00010110	16	SYN
23	00010111	17	ETB
24	00011000	18	CAN
25	00011001	19	EM
26	00011010	1A	SUB
27	00011011	1B	ESC
28	00011100	1C	FS
29	00011101	1D	GS
30	00011110	1E	RS
31	00011111	1F	US
32	00100000	20	SP
33	00100001	21	!
34	00100010	22	"
35	00100011	23	#
36	00100100	24	$
37	00100101	25	%
38	00100110	26	&
39	00100111	27	'
40	00101000	28	(
41	00101001	29	)
42	00101010	2A	*
43	00101011	2B	+
44	00101100	2C	,
45	00101101	2D	−
46	00101110	2E	.
47	00101111	2F	/
48	00110000	30	0
49	00110001	31	1
50	00110010	32	2
51	00110011	33	3
52	00110100	34	4
53	00110101	35	5
54	00110110	36	6
55	00110111	37	7
56	00111000	38	8
57	00111001	39	9
58	00111010	3A	:
59	00111011	3B	;
60	00111100	3C	<
61	00111101	3D	=
62	00111110	3E	>
63	00111111	3F	?

Decimal	Binary	Hexadecimal	ASCII
64	01000000	40	@
65	01000001	41	A
66	01000010	42	B
67	01000011	43	C
68	01000100	44	D
69	01000101	45	E
70	01000110	46	F
71	01000111	47	G
72	01001000	48	H
73	01001001	49	I
74	01001010	4A	J
75	01001011	4B	K
76	01001100	4C	L
77	01001101	4D	M
78	01001110	4E	N
79	01001111	4F	O
80	01010000	50	P
81	01010001	51	Q
82	01010010	52	R
83	01010011	53	S
84	01010100	54	T
85	01010101	55	U
86	01010110	56	V
87	01010111	57	W
88	01011000	58	X
89	01011001	59	Y
90	01011010	5A	Z
91	01011011	5B	[
92	01011100	5C	\
93	01011101	5D	]
94	01011110	5E	^
95	01011111	5F	–
96	01100000	60	'
97	01100001	61	a
98	01100010	62	b
99	01100011	63	c
100	01100100	64	d
101	01100101	65	e
102	01100110	66	f
103	01100111	67	g
104	01101000	68	h
105	01101001	69	i
106	01101010	6A	j
107	01101011	6B	k
108	01101100	6C	l
109	01101101	6D	m
110	01101110	6E	n
111	01101111	6F	o

Decimal	Binary	Hexadecimal	ASCII
112	01110000	70	p
113	01110001	71	q
114	01110010	72	r
115	01110011	73	s
116	01110100	74	t
117	01110101	75	u
118	01110110	76	v
119	01110111	77	w
120	01111000	78	x
121	01111001	79	y
122	01111010	7A	z
123	01111011	7B	{
124	01111100	7C	:
125	01111101	7D	}
126	01111110	7E	
127	01111111	7F	DEL
128	10000000	80	
129	10000001	81	
130	10000010	82	
131	10000011	83	
132	10000100	84	
133	10000101	85	
134	10000110	86	
135	10000111	87	
136	10001000	88	
137	10001001	89	
138	10001010	8A	
139	10001011	8B	
140	10001100	8C	
141	10001101	8D	
142	10001110	8E	
143	10001111	8F	
144	10010000	90	
145	10010001	91	
146	10010010	92	
147	10010011	93	
148	10010100	94	
149	10010101	95	
150	10010110	96	
151	10010111	97	
152	10011000	98	
153	10011001	99	
154	10011010	9A	
155	10011011	9B	
156	10011100	9C	
157	10011101	9D	
158	10011110	9E	
159	10011111	9F	

Decimal	Binary	Hexadecimal
160	10100000	A0
161	10100001	A1
162	10100010	A2
163	10100011	A3
164	10100100	A4
165	10100101	A5
166	10100110	A6
167	10100111	A7
168	10101000	A8
169	10101001	A9
170	10101010	AA
171	10101011	AB
172	10101100	AC
173	10101101	AD
174	10101110	AE
175	10101111	AF
176	10110000	B0
177	10110001	B1
178	10110010	B2
179	10110011	B3
180	10110100	B4
181	10110101	B5
182	10110110	B6
183	10110111	B7
184	10111000	B8
185	10111001	B9
186	10111010	BA
187	10111011	BB
188	10111100	BC
189	10111101	BD
190	10111110	BE
191	10111111	BF
192	11000000	C0
193	11000001	C1
194	11000010	C2
195	11000011	C3
196	11000100	C4
197	11000101	C5
198	11000110	C6
199	11000111	C7
200	11001000	C8
201	11001001	C9
202	11001010	CA
203	11001011	CB
204	11001100	CC
205	11001101	CD
206	11001110	CE
207	11001111	CF

Decimal	Binary	Hexadecimal
208	11010000	D0
209	11010001	D1
210	11010010	D2
211	11010011	D3
212	11010100	D4
213	11010101	D5
214	11010110	D6
215	11010111	D7
216	11011000	D8
217	11011001	D9
218	11011010	DA
219	11011011	DB
220	11011100	DC
221	11011101	DD
222	11011110	DE
223	11011111	DF
224	11100000	E0
225	11100001	E1
226	11100010	E2
227	11100011	E3
228	11100100	E4
229	11100101	E5
230	11100110	E6
231	11100111	E7
232	11101000	E8
233	11101001	E9
234	11101010	EA
235	11101011	EB
236	11101100	EC
237	11101101	ED
238	11101110	EE
239	11101111	EF
240	11110000	F0
241	11110001	F1
242	11110010	F2
243	11110011	F3
244	11110100	F4
245	11110101	F5
246	11110110	F6
247	11110111	F7
248	11111000	F8
249	11111001	F9
250	11111010	FA
251	11111011	FB
252	11111100	FC
253	11111101	FD
254	11111110	FE
255	11111111	FF

Decibels and ratios of power, voltage and current

dB	Power ratio	Voltage/current ratio
−99	1.258925×10^{-10}	1.122018×10^{-5}
−98	1.584893×10^{-10}	1.258925×10^{-5}
−97	1.995262×10^{-10}	1.412538×10^{-5}
−96	2.511887×10^{-10}	1.584893×10^{-5}
−95	3.162278×10^{-10}	1.778279×10^{-5}
−94	3.981072×10^{-10}	1.995262×10^{-5}
−93	5.011873×10^{-10}	2.238721×10^{-5}
−92	6.309573×10^{-10}	2.511886×10^{-5}
−91	7.943282×10^{-10}	2.818383×10^{-5}
−90	$\mathbf{1 \times 10^{-9}}$	$\mathbf{3.162278 \times 10^{-5}}$
−89	1.258925×10^{-9}	3.548134×10^{-5}
−88	1.584893×10^{-9}	3.981072×10^{-5}
−87	1.995262×10^{-9}	4.466836×10^{-5}
−86	2.511886×10^{-9}	5.011872×10^{-5}
−85	3.162278×10^{-9}	5.623413×10^{-5}
−84	3.981072×10^{-9}	6.309574×10^{-5}
−83	5.011872×10^{-9}	7.079458×10^{-5}
−82	6.309573×10^{-9}	7.943282×10^{-5}
−81	7.943282×10^{-9}	8.912510×10^{-5}
−80	$\mathbf{1 \times 10^{-8}}$	$\mathbf{1 \times 10^{-4}}$
−79	1.258925×10^{-8}	1.122018×10^{-4}
−78	1.584893×10^{-8}	1.258925×10^{-4}
−77	1.995262×10^{-8}	1.412538×10^{-4}
−76	2.511887×10^{-8}	1.584893×10^{-4}
−75	3.162278×10^{-8}	1.778279×10^{-4}
−74	3.981072×10^{-8}	1.995262×10^{-4}
−73	5.011873×10^{-8}	2.238721×10^{-4}
−72	6.309573×10^{-8}	2.511886×10^{-4}
−71	7.943282×10^{-8}	2.818383×10^{-4}
−70	$\mathbf{1 \times 10^{-7}}$	$\mathbf{3.162278 \times 10^{-4}}$
−69	1.258925×10^{-7}	3.548134×10^{-4}
−68	1.584893×10^{-7}	3.981072×10^{-4}
−67	1.995262×10^{-7}	4.466836×10^{-4}
−66	2.511887×10^{-7}	5.011872×10^{-4}
−65	3.162278×10^{-7}	5.623413×10^{-4}
−64	3.981072×10^{-7}	6.309574×10^{-4}
−63	5.011872×10^{-7}	7.079458×10^{-4}
−62	6.309573×10^{-7}	7.943282×10^{-4}

dB	Power ratio	Voltage/current ratio
−61	7.943282×10^{-7}	8.912510×10^{-4}
−60	**1×10^{-6}**	**1×10^{-3}**
−59	1.258925×10^{-6}	1.122018×10^{-3}
−58	1.584893×10^{-6}	1.258925×10^{-3}
−57	1.995262×10^{-6}	1.412538×10^{-3}
−56	2.511886×10^{-6}	1.584893×10^{-3}
−55	3.162278×10^{-6}	1.778279×10^{-3}
−54	3.981072×10^{-6}	1.995262×10^{-3}
−53	5.011872×10^{-6}	2.238721×10^{-3}
−52	6.309574×10^{-6}	2.511886×10^{-3}
−51	7.943282×10^{-6}	2.818383×10^{-3}
−50	**1×10^{-5}**	**3.162278×10^{-3}**
−49	1.258925×10^{-5}	3.548134×10^{-3}
−48	1.584893×10^{-5}	3.981072×10^{-3}
−47	1.995262×10^{-5}	4.466836×10^{-3}
−46	2.511886×10^{-5}	5.011872×10^{-3}
−45	3.162278×10^{-5}	5.623413×10^{-3}
−44	3.981072×10^{-5}	6.309574×10^{-3}
−43	5.011872×10^{-5}	7.079458×10^{-3}
−42	6.309574×10^{-5}	7.943282×10^{-3}
−41	7.943282×10^{-5}	8.912509×10^{-3}
−40	**1×10^{-4}**	**0.01**
−39	1.258925×10^{-4}	0.01122018
−38	1.584893×10^{-4}	0.01258925
−37	1.995262×10^{-4}	0.01412538
−36	2.511886×10^{-4}	0.01584893
−35	3.162278×10^{-4}	0.01778279
−34	3.981072×10^{-4}	0.01995262
−33	5.011872×10^{-4}	0.02238721
−32	6.309574×10^{-4}	0.02511887
−31	7.943282×10^{-4}	0.02818383
−30	**1×10^{-3}**	**0.03162277**
−29	1.258925×10^{-3}	0.03548134
−28	1.584893×10^{-3}	0.03981072
−27	1.995262×10^{-3}	0.04466836
−26	2.511886×10^{-3}	0.05011872
−25	3.162278×10^{-3}	0.05623413
−24	3.981072×10^{-3}	0.06309573
−23	5.011872×10^{-3}	0.07079457
−22	6.309574×10^{-3}	0.07943282
−21	7.943282×10^{-3}	0.08912510

dB	Power ratio	Voltage/current ratio
−20	**0.01**	**0.1**
−19	0.01258925	0.1122018
−18	0.01584893	0.1258925
−17	0.01995262	0.1412538
−16	0.02511887	0.1584893
−15	0.03162277	0.1778279
−14	0.03981072	0.1995262
−13	0.05011872	0.2238721
−12	0.06309573	0.2511886
−11	0.07943282	0.2818383
−10	**0.1**	**0.3162278**
−9	0.1258925	0.3548134
−8	0.1584893	0.3981072
−7	0.1995262	0.4466836
−6	0.2511886	0.5011872
−5	0.3162278	0.5623413
−4	0.3981072	0.6309574
−3	0.5011872	0.7079458
−2	0.6309574	0.7943282
−1	0.7943282	0.8912510
0	**1**	**1**
1	1.258925	1.122018
2	1.584893	1.258925
3	1.995262	1.412538
4	2.511886	1.584893
5	3.162278	1.778279
6	3.981072	1.995262
7	5.011872	2.238721
8	6.309574	2.511886
9	7.943282	2.818383
10	**10**	**3.162278**
11	12.58925	3.548134
12	15.84893	3.981072
13	19.95262	4.466836
14	25.11886	5.011872
15	31.62278	5.623413
16	39.81072	6.309573
17	50.11872	7.079457
18	63.09573	7.943282
19	79.43282	8.912510
20	**100**	**10**

dB	Power ratio	Voltage/current ratio
21	125.8925	11.22018
22	158.4893	12.58925
23	199.5262	14.12538
24	251.1886	15.84893
25	316.2278	17.78279
26	398.1072	19.95262
27	501.1872	22.38721
28	630.9574	25.11886
29	794.3282	28.18383
30	$\mathbf{1 \times 10^{-3}}$	**31.62278**
31	1258.925	35.48134
32	1584.893	39.81072
33	1995.262	44.66836
34	2511.886	50.11872
35	3162.278	56.23413
36	3981.072	63.09573
37	5011.873	70.79458
38	6309.573	79.43282
39	7943.282	89.12509
40	**10000**	**100**
41	12589.25	112.2018
42	15848.93	125.8925
43	19952.62	141.2538
44	25118.86	158.4893
45	31622.78	177.8279
46	39810.72	199.5262
47	50118.72	223.8721
48	63095.73	251.1886
49	79432.82	281.8383
50	**100000**	**316.2278**
51	125892.5	354.8134
52	158489.3	398.1072
53	199526.2	446.6836
54	251188.6	501.1872
55	316227.8	562.3413
56	398107.2	630.9573
57	501187.2	707.9458
58	630957.3	794.3282
59	794328.2	891.2509
60	**100000**	**1000**

dB	Power ratio	Voltage/current ratio
61	1258925	1122.018
62	1584893	1258.925
63	1995262	1412.538
64	2511886	1584.893
65	3162278	1778.279
66	3981072	1995.262
67	5011872	2238.721
68	6309573	2511.886
69	7943282	2818.383
70	$\mathbf{1 \times 10^7}$	**3162.278**
71	1.258925×10^7	3548.134
72	1.584893×10^7	3981.072
73	1.995262×10^7	4466.836
74	2.511886×10^7	5011.872
75	3.162278×10^7	5623.413
76	3.981072×10^7	6309.573
77	5.011872×10^7	7079.458
78	6.309574×10^7	7943.282
79	7.943282×10^7	8912.510
80	$\mathbf{1 \times 10^8}$	**10000**
81	1.258925×10^8	11220.18
82	1.584893×10^8	12589.25
83	1.995262×10^8	14125.37
84	2.511886×10^8	15848.93
85	3.162278×10^8	17782.79
86	3.981072×10^8	19952.62
87	5.011872×10^8	22387.21
88	6.309573×10^8	25118.86
89	7.943283×10^8	28183.83
90	$\mathbf{1 \times 10^9}$	**31622.78**
91	1.258925×10^9	35481.34
92	1.584893×10^9	39810.72
93	1.995262×10^9	44668.36
94	2.511886×10^9	50118.72
95	3.162278×10^9	56234.13
96	3.981072×10^9	63095.73
97	5.011872×10^9	70794.58
98	6.309574×10^9	79432.82
99	7.943282×10^9	89125.09
100	$\mathbf{1 \times 10^{10}}$	**100000**

Transmission line power levels and voltages

Level (dBm)	Power (W)	Line voltage (V) $Z_0 = 50$ ohm	$Z_0 = 75$ ohm
−39	1.258×10^{-7}	2.508×10^{-3}	3.072×10^{-3}
−38	1.584×10^{-7}	2.815×10^{-3}	3.447×10^{-3}
−37	1.995×10^{-7}	3.158×10^{-3}	3.868×10^{-3}
−36	2.511×10^{-7}	3.543×10^{-3}	4.340×10^{-3}
−35	3.162×10^{-7}	3.976×10^{-3}	4.870×10^{-3}
−34	3.981×10^{-7}	4.461×10^{-3}	5.464×10^{-3}
−33	5.011×10^{-7}	5.005×10^{-3}	6.130×10^{-3}
−32	6.309×10^{-7}	5.616×10^{-3}	6.879×10^{-3}
−31	7.943×10^{-7}	6.302×10^{-3}	7.718×10^{-3}
−30	$\mathbf{1.000 \times 10^{-6}}$	$\mathbf{7.071 \times 10^{-3}}$	$\mathbf{8.660 \times 10^{-3}}$
−29	1.258×10^{-6}	7.933×10^{-3}	9.716×10^{-3}
−28	1.584×10^{-6}	8.901×10^{-3}	1.090×10^{-2}
−27	1.995×10^{-6}	9.988×10^{-3}	1.223×10^{-2}
−26	2.511×10^{-6}	1.120×10^{-2}	1.372×10^{-2}
−25	3.162×10^{-6}	1.257×10^{-2}	1.540×10^{-2}
−24	3.981×10^{-6}	1.410×10^{-2}	1.727×10^{-2}
−23	5.011×10^{-6}	1.583×10^{-2}	1.938×10^{-2}
−22	6.309×10^{-6}	1.776×10^{-2}	2.175×10^{-2}
−21	7.943×10^{-6}	1.992×10^{-2}	2.440×10^{-2}
−20	$\mathbf{9.999 \times 10^{-5}}$	$\mathbf{2.236 \times 10^{-2}}$	$\mathbf{2.738 \times 10^{-2}}$
−19	1.258×10^{-5}	2.508×10^{-2}	3.072×10^{-2}
−18	1.584×10^{-5}	2.815×10^{-2}	3.447×10^{-2}
−17	1.995×10^{-5}	3.158×10^{-2}	3.868×10^{-2}
−16	2.511×10^{-5}	3.543×10^{-2}	4.340×10^{-2}
−15	3.162×10^{-5}	3.976×10^{-2}	4.870×10^{-2}
−14	3.981×10^{-5}	4.461×10^{-2}	5.464×10^{-2}
−13	5.011×10^{-5}	5.005×10^{-2}	6.130×10^{-2}
−12	6.309×10^{-5}	5.616×10^{-2}	6.879×10^{-2}
−11	7.943×10^{-5}	6.302×10^{-2}	7.718×10^{-2}
−10	$\mathbf{9.999 \times 10^{-5}}$	$\mathbf{7.071 \times 10^{-2}}$	$\mathbf{8.660 \times 10^{-2}}$
−9	1.258×10^{-4}	7.933×10^{-2}	9.716×10^{-1}
−8	1.584×10^{-4}	8.901×10^{-2}	1.090×10^{-1}
−7	1.995×10^{-4}	9.988×10^{-2}	1.223×10^{-1}
−6	2.511×10^{-4}	1.120×10^{-1}	1.372×10^{-1}
−5	3.162×10^{-4}	1.257×10^{-1}	1.540×10^{-1}
−4	3.981×10^{-4}	1.410×10^{-1}	1.727×10^{-1}
−3	5.011×10^{-4}	1.583×10^{-1}	1.938×10^{-1}
−2	6.309×10^{-4}	1.776×10^{-1}	2.175×10^{-1}
−1	7.943×10^{-4}	1.992×10^{-1}	2.440×10^{-1}

Level (dBm)	Power (W)	Line voltage (V)	
		$Z_0 = 50$ ohm	$Z_0 = 75$ ohm
0	$\mathbf{1.000 \times 10^{-3}}$	$\mathbf{2.236 \times 10^{-1}}$	$\mathbf{2.738 \times 10^{-1}}$
1	1.258×10^{-3}	2.508×10^{-1}	3.072×10^{-1}
2	1.584×10^{-3}	2.815×10^{-1}	3.447×10^{-1}
3	1.995×10^{-3}	3.158×10^{-1}	2.868×10^{-1}
4	2.511×10^{-3}	3.543×10^{-1}	4.340×10^{-1}
5	3.162×10^{-3}	3.976×10^{-1}	4.870×10^{-1}
6	3.981×10^{-3}	4.461×10^{-1}	5.464×10^{-1}
7	5.011×10^{-3}	5.005×10^{-1}	6.130×10^{-1}
8	6.309×10^{-3}	5.616×10^{-1}	6.879×10^{-1}
9	7.943×10^{-3}	6.302×10^{-1}	7.718×10^{-1}
10	$\mathbf{9.999 \times 10^{-3}}$	$\mathbf{7.071 \times 10^{-1}}$	$\mathbf{8.660 \times 10^{-1}}$
11	1.258×10^{-2}	7.933×10^{-1}	9.716×10^{-1}
12	1.584×10^{-2}	8.901×10^{-1}	1.090
13	1.995×10^{-2}	9.988×10^{-1}	1.223
14	2.511×10^{-2}	1.120	1.372
15	3.162×10^{-2}	1.257	1.540
16	3.981×10^{-2}	1.410	1.727
17	5.011×10^{-2}	1.583	1.938
18	6.309×10^{-2}	1.776	2.175
19	7.943×10^{-2}	1.992	2.440
20	$\mathbf{1.000 \times 10^{-1}}$	**2.236**	**2.738**
21	1.258×10^{-1}	2.508	3.072
22	1.584×10^{-1}	2.815	3.447
23	1.995×10^{-1}	3.158	3.868
24	2.511×10^{-1}	3.543	4.340
25	3.162×10^{-1}	3.976	4.870
26	3.981×10^{-1}	4.461	5.464
27	5.011×10^{-1}	5.005	6.130
28	6.309×10^{-1}	5.616	6.879
29	7.943×10^{-1}	6.302	7.718
30	**1.000**	**7.071**	**8.660**
31	1.258	7.933	9.716
32	1.584	8.901	1.090×10^1
33	1.995	9.988	1.223×10^1
34	2.511	1.120×10^1	1.373×10^1
35	3.162	1.257×10^1	1.540×10^1
36	3.981	1.410×10^1	1.727×10^1
37	5.011	1.583×10^1	1.938×10^1
38	6.309	1.776×10^1	2.175×10^1
39	7.943	1.992×10^1	2.440×10^1
40	$\mathbf{1.000 \times 10^{-1}}$	$\mathbf{2.236 \times 10^1}$	$\mathbf{2.738 \times 10^1}$

DTMF digits and tone pairs

BCD				Dial digits	Tone pairs	
0	0	0	0	0	941	1336
0	0	0	1	1	697	1209
0	0	1	0	2	697	1336
0	0	1	1	3	697	1477
0	1	0	0	4	770	1209
0	1	0	1	5	770	1336
0	1	1	0	6	770	1477
0	1	1	1	7	852	1209
1	0	0	0	8	852	1336
1	0	0	1	9	852	1477
1	0	1	0	*	941	1209
1	0	1	1	spare (B)	697	1633
1	1	0	0	spare (C)	770	1633
1	1	0	1	spare (D)	852	1633
1	1	1	0	#	941	1477
1	1	1	1	spare (F)	941	1633

Addresses of advisory bodies, standards institutes, and other organisations

American Mobile Telecommunications Association (AMTA)
1150 18th Street NW
Suite 250
Washington DC 20036

American National Standards Institute (ANSI)
1430 Broadway
New York NY 10018

British Standards Institute (BS)
Linford Wood
Milton Keynes
MK14 6LE

Cellular Telecommunications Industry Association (CTIA)
1250 Connecticut Avenue NW
Suite 800
Washington DC 20036

Electronic Industries Association (EIA)
Engineering Department
2001 Eye Street
Washington DC 20006

European Computer Manufacturers Association (ECMA)
11 Rue de Rhone
CH-1204 Geneva
Switzerland

Federal Communications Commission (FCC)
1919 M Street NW
Washington DC 20554

IEEE Communications Society (ComSoc)
305 East 47th Street
New York NY 10017

IEEE Computer Society
1109 Spring Street
Suite 300
Silver Spring
MD 20910

IEEE Standards Board
345 East 47th Street
New York 10017

Information Technology Standards Unit
Department of Trade and Industry
29 Bressenden Place
London
SW1E 5DT

Institute of Electrical and Electronics Engineers (IEEE)
Three Park Avenue 17th Floor
New York NY 10016 5997

Institute of Electrical Engineers (IEE)
Savoy Place
London
WC2R OBL

International Standards Organisation (ISO)
1 Rue de Varembe
CH-1211
Geneva
Switzerland

International Telecommunications Union (ITU)
Place des Nations
1121 Geneva 2
Switzerland

ITU-T
Place des Nations
CH-1211 Geneva 20
Switzerland

National Bureau of Standards (NBS)
Technical Information and Publications Division
Washington DC 20234

National Cable Television Association (NCTA)
1724 Massachusetts Avenue NW
Washington DC 20036

National Computing Centre (NCC)
Oxford Road
Manchester
M1 7ED

Personal Communications Industry Association (PCIA)
500 Montgomery Street
Suite 700
Alexandria VA 22314

Telecommunications Industry Association (TIA)
2500 Wilson Boulevard Suite 300
Arlington VA 22201-3834

US Department of Commerce
National Technical Information Service
5285 Port Royal Road
Springfield
VA 22161

Basic logic gates

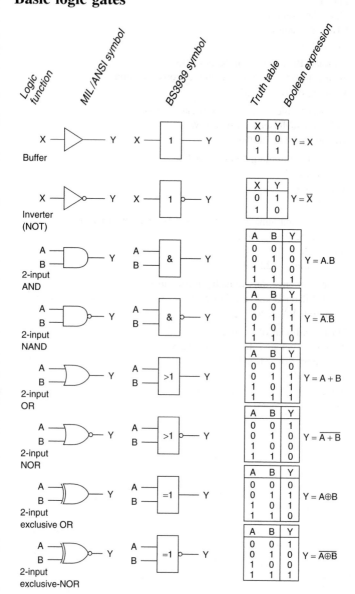

Index